Reuse in Intelligent Systems

Editors

Stuart H Rubin
Space and Naval Warfare Systems Center
San Diego, USA

Lydia Bouzar-Benlabiod
Ecole Nationale Supérieure d'Informatique
Algiers, Algeria

CRC Press
Taylor & Francis Group
Boca Raton London New York

CRC Press is an imprint of the
Taylor & Francis Group, an **informa** business

A SCIENCE PUBLISHERS BOOK

CRC Press
Taylor & Francis Group
6000 Broken Sound Parkway NW, Suite 300
Boca Raton, FL 33487-2742

First issued in paperback 2021

Version Date: 20200211

ISBN 13: 978-0-367-51007-7 (pbk)
ISBN 13: 978-0-367-47338-9 (hbk)

Visit the Taylor & Francis Web site at
http://www.taylorandfrancis.com

and the CRC Press Web site at
http://www.crcpress.com

Preface

The aim of this book is to present recent works covering various aspects of reuse in intelligent systems—including Scientific Theory and Technology-Based Applications. New data analytic algorithms, technologies, and tools are sought to be able to manage, integrate, and utilize large amounts of data despite hardware, software, and/or bandwidth constraints; to construct models yielding important data insights; and, to create visualizations to aid in presenting and understanding the data. System development and integration needs to also adapt to these new algorithms, technologies, tools, and needs.

The growth of big data, in part due to it's ubiquity, has increased the need for applying machine learning to solve real-world problems. Besides the large size inherent in big data, these datasets are adversely affected by class imbalance, which contributes to poor machine learning performance. *Experimental Studies on the Impact of Data Sampling with Severely Imbalanced Big Data* demonstrates the efficacy of machine learning classification with big data when confronted with the class imbalance problem. Two case studies with diverse range of class ratios between majority and minority classes, across various levels of class imbalance, have been provided. In the first case study, we process four big balanced datasets and artificially generate five imbalanced big datasets from the original full datasets, with target minority classes of 10%, 1%, 0.1%, 0.01%, and 0.001%. Random undersampling is then applied to balance the binary class in each of the generated imbalanced datasets to 50:50 class ratios. All machine learning models were built using the Random Forest classifier. For the second case study, a real-world Medicare fraud detection problem is introduced, which focuses on applying various random undersampling class ratios and injecting additional artificial class imbalance. Three learners (Logistic Regression, Random Forest, Gradient Boosted Trees) were employed. The results show that, in terms of class imbalanced data, ratios from 0.1% to 1.0% of the minority class provide adequate performance even when compared to 10% or even 100% of the original full balanced dataset. Furthermore, a balanced random undersampling ratio, when applied to the imbalanced big dataset, led to similarities in the average performance when compared to using the entire big dataset. Also, when the minority class is severely imbalanced, the balanced class ratio is not always the best option with slightly more imbalance, such as 10% or even 1%, providing better overall model performance.

The US healthcare industry produces copious amounts of big data, which includes information such as patient records and provider claims. Leveraging this big data is becoming increasingly important in keeping healthcare programs affordable and maintaining high levels of medical care— especially for the rising elderly population. The elderly are experiencing increased life expectancy, with continuing healthcare needs later in life and the need for programs, such as US Medicare, to help with associated medical expenses. Unfortunately, due to healthcare frauds, these programs are being adversely affected, draining resources and reducing quality and accessibility of necessary healthcare services. The detection of fraud is critical in being able to identify and subsequently stop these perpetrators. The application of machine learning methods to big data can be leveraged to improve current fraud detection processes and reduce the resources needed to investigate possible fraudulent activities. *Big Data and Class Imbalance in Medicare Fraud Detection* presents two case studies for detecting fraud across several big Medicare claims datasets from 2012 to 2015, considering the severe class imbalance between fraud and non-fraud claims, with actual fraud labels from the List of Excluded Individuals/Entities (LEIE). The first case study employs the Random Forest model with random undersampling, to mitigate some of the adverse effects of class imbalance and to generate seven different class distributions for a comparison of performance results with Medicare Part B data. The second case study expands upon the first by taking the best class distribution from the first and providing results for two additional Medicare datasets and a combined dataset. We demonstrate that 90:10 is the best class distribution; whereas, the balanced and two of the highly imbalanced distributions produced the worst fraud detection performance. Furthermore, we show that the commonly used ratio of 50:50 (balanced) was not significantly better than using a 99:1 (imbalanced) class distribution. The study clearly demonstrates the need to apply at least some sampling to big data with class imbalance and suggests the 50:50 class distribution does not produce the best Medicare fraud detection results.

Researchers and practitioners commonly use feature selection and data sampling to counter high dimensionality and class imbalance. *How to Optimally Combine Univariate and Multivariate Feature Selection with Data Sampling for Classifying Noisy, High Dimensional and Class Imbalanced DNA Microarray Data* was conducted to give practitioners guidance on best practices when analyzing bioinformatics data that exhibit both high dimensionality and class imbalance in the context of data noise. Three approaches for combining feature selection and data sampling are compared: (1) data sampling followed by feature selection with the training data being built using the selected features and the unsampled data; (2) data sampling followed by feature selection with the training data being built using the selected features and the sampled data, and (3) feature selection followed by data sampling with the training data being built using the selected features and the sampled data. Additionally, the importance of alleviating class imbalance is investigated (by applying data sampling) for classification problems on bioinformatics datasets. We explored three major forms of feature selection (feature rankers, filter-based subset selection, and wrapper-based subset selection), as well as a commonly used data sampling technique (Random Undersampling). All experiments were conducted using ten gene expression datasets which were first determined to be relatively free of noise. Then, noise is artificially injected, creating three levels of data quality to simulate real-world scenarios. Final models are built using six different classification algorithms. Empirical results show that the best performing approach is feature selection followed by data sampling, across all data quality levels. We also show that alleviating class imbalance (e.g., by applying Random Undersampling), in conjunction with reducing high dimensionality, will achieve improved classification performance for bioinformatics classification problems compared to reducing the high dimensionality without alleviating the class imbalance.

Given the number of new movies being released every week, online recommenders play a significant role in suggesting movies for individuals or groups of people to watch—either at home or at movie theaters. Making recommendations relevant to the interests of an individual, however, is not a trivial task due to the diversity in individual preferences. To address this issue *Movie Recommendations Based on the Recurrent Neural Network Model* introduces a novel movie recommender system that suggests movies appealing (to a certain degree) to movie goers. Recommendation systems are an important part of suggesting movies—especially in streaming services. For streaming movie services like Netflix, recommendation systems are essential for helping users find new movies to view. In this paper, we propose a deep learning approach based on autoencoders to produce a collaborative filtering system, which predicts movie ratings for a user based on a large database of ratings from other users. Using the MovieLens dataset, we explore the use of deep learning to predict users' ratings on new movies, thereby enabling movie recommendations. To verify the novelty and accuracy of our deep learning approach, we compare our approach to standard collaborative filtering techniques: k-nearest neighbor and matrix factorization. The experimental results show that our recommendation system outperforms a user-based neighborhood baseline in terms of root mean squared error on predicted ratings. In addition, we have conducted other user studies, which were straightly based on human assessment, on movie recommendations made by on our recommender, along with Amazon and Redbox, two well-known movie recommenders. Performance of recommenders are compared and the empirical study further verified the merit and novelty of our movie recommender. The design the new recommender, which is simple and domain-independent, can easily be extended to make recommendations on items other than movies.

The popularization of MOOCs in recent years has consolidated this learning format in the open education scenario, with the emergence of new providers, new available courses, and more universities becoming partners. However, this accelerated expansion makes it difficult for students to find the most appropriate content; and, some recommendation systems have emerged to support such decisions. A Recommendation System Enhanced by Topic Modeling for Knowledge Reuse in MOOCs Ecosystems advances in the investigation of MOOCs recommendation systems, addressing the use of Linked Open Data, enhanced by topics modeling and labeling methods to integrate and reuse data. Moreover, this chapter applies the concepts of software ecosystem (SECO) in the modeling of MOOCs ecosystems, identifying interactions and benefits of this approach. Finally, an example of use is conducted to verify usability and how the techniques perform to recommend courses (or parts of courses) in multiple MOOCs providers.

Petri nets (PN) are a mathematical tool that allows for complex algorithms to be modeled visually and demonstrates the capabilities to quickly observe the outcomes of an algorithm. With additional variables, in our case time, the standard PN model can be enhanced to give greater modeling capabilities to a developer. *Towards a Computer Vision Based Approach for Developing Algorithms for Soccer Playing Robots* focuses on the actions a robotic goalkeeper should take in a soccer match, a Timed Petri net (TdPN) was used to model and simulate the system. The TdPN was developed to take inputs from machine learning models, which includes the detection of the soccer ball and other robots as well as the distances to each from pictures taken by the robot. Using these predictors, an initial marking for the TdPN can be determined, which when simulated will choose the desirable action based on the input stimuli of what the robotic goalkeeper sees. Additionally, we analyze our TdPN to see where the model can be modified and/or expanded to account for changes in the future.

Verification tools for hybrid systems with mixed discrete-continuous behavior are becoming more and more powerful, but their applicability to high-dimensional models is still restricted. *Context-dependent Reachability Analysis for Hybrid Systems* proposes an improvement for a certain class of verification techniques based on flow-pipe construction. In previous work we presented a method that allows for decomposition of the state space of a hybrid system, such that the analysis can be done in sub-spaces of lower dimensions, instead of the global high-dimensional space. In this paper, we present an approach to construct such decompositions automatically, to analyze the dynamics in each of the sub-spaces, and to select for each sub-space an individual well-suited verification method. Our experimental evaluation demonstrates the general applicability of our approach and shows a remarkable speed-up on decomposable systems with heterogeneous dynamics.

Attackers can leverage several techniques to compromise computer networks—ranging from sophisticated malware to Distributed Denial of Service (DDoS) attacks that target the application layer. Application layer DDoS attacks, such as Slow Read, are implemented with just enough traffic to tie up CPU or memory resources causing web and application servers to go offline. Such attacks can mimic legitimate network requests making them difficult to detect. *Netflow Feature Evaluation for the Detection of Slow Read HTTP Attacks* explores eight machine learners for detecting Slow Read DDoS attacks on web servers at the application layer. Our approach uses a generated dataset based upon Netflow data collected at the application layer on a live network environment. Our generated dataset consists of real-world network data collected from a production network. The eight machine learners provide us with a more comprehensive analysis of Slow Read detection models. It is essential to know which features reflect the most significant value regarding the learners' performance. Selective feature evaluation has several methods used to specify the attribute evaluator and search methods. Correlation Feature Selection (CFS) evaluates the worth of a subset of attributes by considering the individual predictive ability of each feature. In machine learning and statistics, feature selection methods such as single-attribute, subset attributes selection, and Principal Component Analysis (PCA) are excellent approaches for choosing a subset of relevant features for enhancing machine learning models. We explore the use of these methods to improve the machine learners for detecting Slow Read DDoS attacks on web servers at the application layer. Experimental results show that the machine learners were successful in identifying the Slow Read attacks with a high detection and low false alarm rate. The experiment demonstrates that our chosen Netflow features and feature selection methods are discriminative enough to detect such attacks with 90 percent accuracy.

Server Logs are an important source of information for diagnosing abnormal behavior as well as proactive error handling. Generally, errors are examined manually by human experts, which takes a considerable amount of time and effort to prevent the system from failure. The system log files, besides other attributes such as time and location, and contains messages in textual form, which is essential for analyzing behavior logs and understanding the cause of errors. *Predictive Analysis of Server Log Data for Forecasting Events* forecasts future server events, which helps the data analyst to predict future system failure. We are reusing the sequence of events and forecasting future events for abnormal behavior detection by the system. Accurate forecasting of time-series events is optimum for active strategies, excellent performance of the system, preventive maintenance, and complete shut-down. We have explored the LSTM (Long short-term memory) algorithm, Holt-Winters, and ARIMA algorithms and compared the results. We found that LSTM produces promising results for forecasting future events.

Stuart H Rubin
Lydia Bouzar-Benlabiod

Contents

Chapter **1**

Experimental Studies on the Impact of Data Sampling with Severely Imbalanced Big Data

Tawfiq Hasanin, Taghi M Khoshgoftaar* and Richard A Bauder

1. Introduction

Recent developments in technology have caused the growth of raw data to occur at an explosive rate. This has resulted in immense opportunity for knowledge discovery and data engineering research to play an essential role in a wide range of applications from enterprise information processing to governmental decision-making support systems, and microscale data analysis to macroscale knowledge discovery.

In defining the term "big data", scholars provide many examples throughout the literature [1]–[4]. In general, big data refers to large and complex data, made up of structured and unstructured data which are too big, or too computationally expensive, to be managed by traditional data mining and *Machine Learning* (ML) techniques. Today, a huge amount of information and data are stored in digital mediums which make it easier to use more advanced methods to extract meaningful information. The general consensus is that there are certain attributes that characterize big data. Throughout the literature, the task of defining big data has proven rather complicated, without a universally accepted definition [5]. Recently, Senthilkumar et al. [5]

Department of Computer & Electrical Engineering and Computer Science, Florida Atlantic University, Boca Raton, Florida, USA.
Emails: thasanin2013@fau.edu; rbauder2014@fau.edu
* Corresponding author: khoshgof@fau.edu

provided a definition specifically for healthcare, categorizing big data into six V's: Volume, Variety, Velocity, Veracity, Variability, and Value. Volume refers to vast quantities of data, Variety applies to high levels of complexity of data (i.e., incorporating data from different sources, mash-ups), Velocity represents the high frequency at which new data is generated/collected, Veracity pertains to the correctness of the data, Variability refers to sizable fluctuations, or variation, in the data, and Value signifies significant data quality in reference to the intended results (e.g., fraud detection).

When focusing on learning from the available information, ML is a branch of *Artificial Intelligence* (AI) that studies the ability to learn without explicitly being programmed to do so. Compared to more traditional solutions, ML algorithms generally provide good results [6]–[8]. Moreover, the traditional techniques cannot cope with such large amounts of data, thus fueling a growing need to run machine learning and data mining methods on increasingly larger datasets [9]. With this kind of growth in dataset size, many problems surface. An important problem that affects learning from big data is class imbalance. Class imbalance refers to the condition where the classes are not represented equally [10]–[12]. Generally speaking, most labeled datasets have some inequality in the number of classes, such as having very few fraud cases relative to the non-fraud cases. The vast majority of instances belong to one or several classes and a very small minority belong to the class, or classes, of interest. Some real-world examples of minority class, or classes of interest, include positive cancer diagnoses, medical fraud cases, and airport security breaches. All of these minority classes are significantly less likely to occur in comparison to the normative situation. Thus, the abnormal cases (minority or positive classes) are the ones we want to successfully detect and deal with accordingly.

The problem of imbalanced learning has attracted a significant amount of attention from academia, industry, and government agencies in recent years [13]. An issue with using imbalanced training data is how it might impact the performance of a ML algorithm, which assumes balanced class distributions or equal misclassification costs [13], [14]. For that reason, when presented with imbalanced datasets, these algorithms usually fail to properly represent the characteristics of the data and perform poorly in correctly classifying the data [13], [15]. There are many different techniques when dealing with class imbalanced datasets, which include the following:

- Collect more data for the minority class. However, this can be difficult to achieve since the minority class can be hard to collect or unavailable, as seen with Medicare fraud.
- Apply several classification algorithms to assess which one performs better on a particular dataset.

- Use sampling methods on the data to lessen the impact of class imbalance. Sampling generates new datasets from the original. There are many sampling methods such as *Random Oversampling* (ROS) [16], *Random Undersampling* (RUS) [16], the *Synthetic Minority Over-sampling TEchnique* (SMOTE) [17] and cost-sensitive learning [18].

- Lastly, though not specifically an imbalance solution, the use of different performance metrics to give additional insights into model performance can help in assessing the true impact of class imbalance. For example, accuracy can be very misleading in reporting performance on imbalanced datasets. Accuracy, sometimes called error rate, usually applies a 0.50 threshold to decide between classes, and this is typically incorrect with severely class imbalanced datasets. In Section 3.3, we explain the Area Under the Receiver Operating Characteristic (ROC) Curve (AUC) which can be used in lieu of accuracy. AUC is an average performance of all operating points on the ROC curve for a particular learner.

In this study, we focus on sampling to reduce the impact of class imbalance on machine learning models. Sampling techniques usually fall into two categories: undersampling the majority class or oversampling the minority class. The first removes instances from the majority, while the latter adds instances to the minority class. RUS is based on randomly removing instances from the majority class, but other methods selectively undersample the majority class, while keeping the original population of the minority class [19]. ROS randomly oversamples the minority class. SMOTE oversamples the minority class by creating "synthetic" examples rather than by oversampling with replacement. For our paper, we apply a RUS-based class imbalance methodology. In general, we did not use ROS to avoid increasing the size of the already large datasets, which can lead to an increase in computation time and expense. Additionally, we avoided using SMOTE because it can create samples that are not real or representative of the actual data, thus misleading ML models [20].

We demonstrate that classification performance across several imbalanced big datasets across different application domains can be significantly improved using RUS without substantially altering the composition of the original data. Our results indicate that having some data imbalance, from 0.1% to 1.0% of the minority class, provides good performance versus using the original imbalanced dataset or a heavily altered balanced dataset. For big data models, the use of RUS implies less loss of information in the negative class, and thus a better overall representation of the original data (unlike the 50:50 class distribution). The following sequence summarizes our approach for the first case study:

- Collect balanced big datasets.
- By simulation, randomly discard positive class instances, generating five different class ratios.
- Apply RUS to each dataset to get a 50:50 (balanced) class distribution.
- Employ *Random Forest* (RF) on each dataset (to include the original, full dataset) to assess classification performance.

In addition to the aforementioned experiments on class imbalance, we also introduce a class imbalance problem using real-world Medicare fraud datasets, as a second case study. From these datasets, we randomly select baskets from the minority (positive) class of 200, 100, and 50 instances. We then apply RUS on each basket, as well as the original full datasets, producing 50:50, 75:25, 65:35, 90:10, and 99:1 class distributions. Model performance is assessed using *Logistic Regression* (LR), RF, and *Gradient Boosted Trees* (GBT) across the different datasets. Our results indicate that applying RUS with more imbalanced class ratios, such as 90:10 and 99:1, provide better performance than the typical 50:50 class ratio.

Our contribution involves clearly demonstrating the adverse impact of class imbalance in big data on machine learning model performance, and lessening these effects by applying RUS and state-of-the-art big data tools and frameworks. More specifically, we present two experiments. In the first, we compare the original datasets, as a baseline, against datasets with balanced and imbalanced class distributions. This helps to determine a good class distribution when using imbalanced big data. In the second experiment, we employ real-world imbalanced Medicare datasets. We create additional severely imbalanced datasets and apply RUS to each. To the best of our knowledge, our work is unique in generating both imbalanced and balanced datasets to determine the effects of class imbalance on big data.

The remainder of this paper is organized as follows. Section 2 provides an overview of related works. Section 3 describes the ML classification algorithms and libraries used in this paper, to include the evaluation strategy with validation techniques and performance metrics. Section 4 introduces our first experiment to include the datasets and how they were processed, model training, and performance evaluation. Section 5 presents our second experiment which involves a real-world Medicare fraud problem, with severe class imbalance. Section 6 presents our conclusions and future work.

2. Related Works

There are several studies that offer a good overview for the problem of imbalanced data which include works such as [13], [21]–[23]. Overall, approaches for addressing the problem of class imbalance fall largely in two groups: data sampling solutions [16], [17], which modify the original training set, and algorithmic modifications [24], which modify existing algorithms trying to benefit from the classification of the minority class. Cost-sensitive solutions [25], [26] combine the two previous options trying to minimize misclassification costs, which are higher for the instances of the minority class.

The researchers in [27] have proposed an enhanced SMOTE algorithm for classification of imbalanced big data using RF. In their work, they introduced a method to work on multi-class imbalanced data. The initial step decomposed the original dataset into subsets of binary classes. The authors then applied the SMOTE algorithm to each subset of imbalanced binary class in order to create balanced data. The results showed that their proposed method outperforms other methods.

Another work that has come to our attention is [28]. Their method won the ECBDL'14 big data challenge for a bioinformatics big data problem. This algorithm, named ROSEFW-RF, is based on several approaches to balance the class distributions through ROS, detecting the most relevant features via an evolutionary feature weighting process and a threshold to choose them, building an appropriate RF model from the pre-processed data, and classifying the test data. From their analysis, they concluded that their approach is very suitable to tackle large-scale bioinformatics classification problems.

A recent study [29] addressed the fact that existing solutions typically follow a divide-and-conquer approach in which the data is split into several chunks that are addressed individually. Next, the partial knowledge acquired from every slice of data is aggregated in multiple ways to solve the entire problem. However, these approaches are missing a global view of the data as a whole, which may result in less accurate models. In their work, the researchers carried out a first attempt on the design of a global evolutionary undersampling model for imbalanced classification problems. These are characterized by having a highly skewed distribution of classes in which evolutionary models are being used to balance the dataset by selecting only the most relevant data. Using Apache Spark [30], they introduced a number of variations to the well-known CHC [31] algorithm to work with very large chromosomes and reduce the costs associated with the fitness evaluation. They discussed some preliminary results, showing the potential of this new kind of evolutionary big data model.

The work in [32] analyzed the performance of several techniques used to deal with imbalanced datasets in big data. The work adopted oversampling, undersampling, and cost-sensitive learning to correctly identify the underrepresented class. An experimental study was carried out to evaluate the performance of the diverse algorithms considered. The results indicated that there is not one approach to imbalanced big data classification that outperforms the others for all the data considered, when using RF.

Another study [33] analyzed the performance of oversampling and undersampling with the decision tree [34] learner. Their work shows that using decision trees with undersampling establishes a reasonable standard for algorithmic comparison. But, it is recommended that the least cost classifier be part of that standard as it can be better than undersampling for relatively modest costs. Oversampling, however, shows little sensitivity. The authors note that there is often little difference in performance when misclassification costs are changed.

One study by our research group [35] discusses four Medicare datasets, and provides an exploratory analysis of fraud detection. They achieved good fraud detection performance particularly for LR and RF. Nevertheless, this study did not include data sampling methods to address the issue of class imbalance. Another recent paper [36] employed undersampling to study the impact of class imbalance by creating four class ratios (80:20, 75:25, 65:35, and 50:50). Using RF and LR learners, the research concluded that the 80:20 class distribution performed the best with low false negative rates.

3. Background

In this section, we describe the machine learning models used in our study. Additionally, we discuss two machine learning frameworks used in our experiments to process and build models with the big datasets. Note that during model training, we keep the default model parameters unless otherwise stated. For the first case study, we maintain configurations that are as similar as possible between the Spark and H$_2$O frameworks. However, for the second case study, we use only the Spark framework for building models on the big Medicare datasets.

3.1 Machine Learning Algorithms

The decision tree is a greedy algorithm that performs a recursive binary partitioning of the features. The tree predicts the label for each bottom-most (leaf) partition. Each partition is chosen greedily by selecting the best split from a set of possible splits, in order to maximize the *information gain* (IG) at a tree node [30]. The node impurity is a measure of the homogeneity of the labels at the node. The current implementation in Spark provides two impurity measures for classification. That is, Gini impurity which is defined by the formula $\sum_{i=1}^{C} f_i(1 - f_i)$ and entropy which is defined by $\sum_{i=1}^{C} -f_i \log f_i$, where f_i is the frequency of label i at the specific node and C is the number of unique labels. The information gain is the difference between the parent node impurity and the weighted sum of the two child node impurities. Note that DT is not directly used in our study, but is an integral part of both the Random Forest and Gradient-Boosted Tree models.

RF [37] is an ensemble approach that can also be thought of as a form of nearest neighbor predictor. Ensembles are a divide-and-conquer approach used to improve performance. The main principle behind ensemble methods is that a group of "weak learners" can come together to form a "strong learner." RF employs the Decision Tree (DT) algorithm as the "weak learner" in the ensemble. In a DT, each branch of the tree represents a feature in the data which divides the instances into more branches based on the values which that feature can take. Information Gain is used to decide the hierarchy of features in the final tree structure. The leaves of the tree represent the final class label. In our study, we applied RF as a base machine learning algorithm. Ensembles have demonstrated good behavior when confronted with imbalanced datasets [38], and it is believed that using one of them as a basis for the comparison should not bias the results regarding the minority class [32]. It is also believed that combining random sub-sampling with RF may overcome the imbalance problem [39]. A recent study [40] used 121 datasets from the *University of California, Irvine* (UCI)[1] Machine Learning Repository to develop a comparison of 179 classifiers arising from 17 families. The study excluded large-scale problems. Their conclusion was that the classifier most likely to be the best was RF.

GBT iteratively trains a sequence of decision trees. On each iteration, the algorithm uses the current ensemble to predict the label of each training instance and then compares the prediction with the true label. The dataset is re-labeled to put more emphasis on training instances with poor predictions. Thus, in the next iteration, the decision tree will help correct previous mistakes. The specific mechanism for re-labeling instances is defined by a loss function. With each iteration, GBT further reduces this loss function on the training data [30].

This model measures the relationship between the categorical dependent variable and one or more independent variables by estimating probabilities using a logistic function, which is the cumulative logistic distribution. Thus, it treats the same set of problems as probit regression using similar techniques, with the latter using a cumulative normal distribution curve instead with the loss function in the formulation given by the logistic loss:

$$L\,(w : x,\, y) := \log(1 + e^{-yw^T x}) \tag{1}$$

---

[1] http://archive.ics.uci.edu/ml/index.php.

For binary classification problems, the algorithm outputs a binary LR model. Given a new data point, denoted by x, the model makes predictions by applying the logistic function $f(z) = \frac{1}{1+e^z}$ where $z = w^T x$. If $(w^T x) > 0.5$, the outcome is positive class, or negative otherwise.

3.2 Machine Learning Frameworks

To ease the process of using ML, engineers build the algorithms within software modules or packages, making sure that they work reliably, quickly, and at-scale. Furthermore, these frameworks are specifically designed to leverage distributed compute resources for processing extremely large datasets. In the context of ML, our work employs two state-of-the-art big data frameworks:

- Apache Spark [30], also referred to as Spark in this paper, provides dramatically increased data processing speed compared to traditional methods and is considered one of the largest big data open source projects [41].

- H_2O is another open source framework that provides a parallel processing engine, analytics, math, and machine learning libraries, along with data preprocessing and evaluation tools. Additionally, it offers a web-based user interface, making learning tasks more accessible to analysts and statisticians who may not have strong programming backgrounds [42].

3.3 Evaluation Strategy

Typically, when training and validating models, datasets are split into two thirds for model training and one third for testing. However, this method has some disadvantages, mainly because only part of the data is in either the training or validating process but not both. To overcome this problem, leave-one-out validation methods help by using the entire dataset in the evaluation process. Two validation methods employing this idea were used in this work, to include k-fold cross-validation and Out-of-Bag error.

In ML, one of the commonly used model evaluation methods is cross-validation (CV), in which a portion of the data is used to train the model while the remaining data is used to validate the built model. K-fold CV is also known as rotation estimation, which evaluates predictive models by partitioning the original sample into several sets of approximately equal size. As seen in Figure 1, the model is trained and tested k times, where each time it is trained on $k-1$ folds and tested on the remaining fold. This is to ensure that all data are used in the classification. A slight modification in the k-fold CV technique is made for some classification problems. With imbalanced data, one typically uses stratified k-fold CV, in which minority and the majority classes

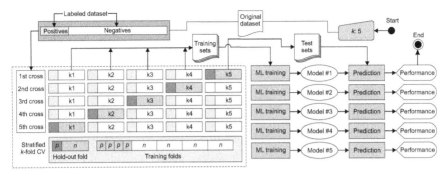

Fig. 1: K-Fold cross validation.

have roughly the same proportions of class labels in each fold, to ensure that minority classes have approximately balanced distribution between training and test sets. When compared to regular CV, the stratification scheme is generally better in terms of bias and variance [43].

A RF model, however, has the ability to internally estimate the performance during run time. This method is called *Out-of-Bag* (OOB) error. While using the entire original dataset, each tree in the forest is built using a different bootstrap sample. Typically, one-third of the dataset is left out and not involved in the current tree construction. This set is used to validate the tree built on the remaining two-thirds, with every other tree in the forest similarly treated. At the end of the run, j is taken as the OOB, the class that received most of the votes every time out of n cases. The proportion of times that j is not equal to the true class of n, averaged over all cases, is the OOB error estimate. The OOB is used, while adding trees to a forest, to achieve a running unbiased estimate of the classification error.[2]

3.4 Area Under ROC Curve

Although the accuracy metric threshold may use values other than the default 0.5 to distinguish between binary classes, the accuracy metric is based on a simple count of the errors which can easily hide information due to class confusion. To measure model performance, we use the *Area Under the Receiver Operating Characteristic Curve* (AUC) metric. AUC is preferred over accuracy as an alternative method for evaluating a classification algorithm [44]. The use of AUC allows us to focus on a classifier's ability to avoid false classification [45]. This is particularly important when working with imbalanced datasets when the positive class, the class of interest, is in the minority.

---

[2] https://www.stat.berkeley.edu/~breiman/RandomForests/cc_home.htm.

3.5 Average and Weighted AUC

In Spark, the average AUC over all of the 5-fold CV folds is used. However, this approach produces different results using H_2O. Scoring the holdout predictions can result in different metrics versus taking the average over all of the 5-fold CV folds. For example, if the sizes of the holdout folds differ significantly, then the average should be replaced with a weighted average. Also, if the CV models map to slightly different probability spaces, which can happen for some models that converge to different local minima, then the confused rank ordering of the combined predictions can lead to a significantly different AUC than the average.[3] Besides the average AUC scores, we retain all of the individual AUC scores for statistical analysis.

3.6 Significance Testing

In order to provide additional rigor around our AUC performance results, we use hypothesis testing to show the statistical significance of the model performance results. Both ANalysis Of VAriance (ANOVA) [46] and post hoc analysis via Tukey's Honestly Significant Different (HSD) [47] tests are used in our study. ANOVA is a statistical test determining whether the means of several groups (or factors) are equal. Tukey's HSD test determines factor means that are significantly different from each other. This test compares all possible pairs of means using a method similar to a t-test, where statistically significant differences are grouped by assigning different letter combinations (e.g., group 'a' is significantly different than group 'b').

3.7 The Problem of Randomization

Randomization can be problematic when balancing datasets by either undersampling or oversampling. Additionally, problems from randomization can adversely affect the resampling of the k-folds in CV, where it randomly divides the data into these k folds. Statistically, having a sample is only part of the population, where the numerical value of a statistic cannot be expected to provide the exact value of the population for any given sample. With RUS, the split is completely random and retains only a fraction of the data. Thus, due to this randomness, RUS performs splits that can be considered lucky or unlucky. Random splits may generate very good (clean) sampled instances (that could increase model performance) or may retain poor and/or noisy instances which may degrade the training process and model performance. Lastly, some algorithms, such as RF, have inherent randomness, whereas others output results where the order of instances is changed. One way to reduce some of

---

[3] https://h2o-release.s3.amazonaws.com/.

the potential negative effects of randomness is by using repetitive methods [48]. We use five repeats in the first case study for each RUS split. This will provide five models for each ratio, which means that we created a total of 200 new datasets from the original four datasets and 4,488 models. Each of these five repetition results were averaged to get the overall model performance. In the second case study, we decided to repeat it 10 times due to the severe class imbalance and relative size of each dataset, which generated 14,400 models.

4. Simulated Imbalanced Data Case Study

For our first case study, we considered four diverse and publicly available big datasets, three of which were gathered from UCI Machine Learning Repository [49]. Because it is likely the most developed branch of learning from imbalanced data [50], binary classification problems are very important in our current research. Hence, our primary focus is on using high-dimensional datasets with binary labels for classification.

The datasets in this case study are listed in Table 1 which include HIGGS, SUSY [51], HEPMASS [52], and sentiment140 [53]. HIGGS, HEPMASS, and SUSY are similar to each other in nature while sentiment140 has a different representation characteristic. More specifically, sentiment140 has a word vector representation [54] with features representing the words and the instances representing tweets. Each value in the data holds either 1, indicating the word exists in the document or 0, indicating the absence of the specific word. Table 1 presents the dimensions, number of class instances, number of features, the learning difficulty, and the performance. The datasets are ordered by level of difficulty. The importance of understanding the difficulty of a dataset is needed because the evaluation result is dependent on the dataset itself, e.g., how difficult it is to learn patterns in a particular dataset. The listed performance is based on building and evaluating RF models on the original, full datasets (with no sampling). There is not a lot of information regarding the learning difficulty of the collected datasets, except from comparing model performance for each dataset in Section 4.4. As the reader will see in Section 6, as well as in Table 3, the performance of each dataset varies and little can be done to significantly improve the model performance with data

Table 1: Datasets.

Dataset	Features	Instances	Difficulty	AUC Performance
HEPMASS	28	10,500,000	Easy	0.953
SUSY	18	5,000,000	Medium	0.874
HIGGS	28	11,000,000	Hard	0.822
sentiment140	109,735	1,600,000	Very hard	0.785

sampling when the dataset is very difficult to learn. In general, the level of difficulty of a dataset may depend on one or more factors, such as the algorithm used, level of noise, or high dimensionality. Readers may refer to [55]–[57] to learn more about level of difficulty.

As stated, the primary purpose for using big balanced datasets is to study the effects of learning with imbalanced data. Usually, methods are used to provide a balanced distribution by modifying imbalanced datasets; however, our method aims to intentionally inject imbalance to provide a general insight about imbalanced datasets. Studies have shown that, in comparison to imbalanced data, a balanced dataset produces an improvement in the overall performance for many classifiers [13].

4.1 The Imbalanced Datasets

Each one of the four datasets is represented with the symbol α next to its name to indicate that it is the original balanced dataset. Five new imbalanced datasets (referred to as data1, data2, data3, data4, and data5, respectively) are created out of the original α balanced datasets with different ratios randomly undersampled from only the positive class.

In binary classification, the idea is to define each instance with a label, either positive or negative. In real-world problems, the positive class is typically more important and is what we want to predict or detect. To mimic real-word problems, we are going to decrease the positive class by randomly discarding them. Table 2(a) shows twenty new imbalanced datasets plus the four original. Typically, RUS removes data from the original dataset. In particular, it randomly selects a set of majority class instances and removes these samples to adjust the balance of the original dataset. Nevertheless, in our experiment, we used undersampling to inject imbalance into the big datasets by removing instances from the positive class. Due to the design of our experiment, we started by randomly sampling the data keeping 10% of the positive class while discarding 90%, thus obtaining a ratio of 10:1 which we call number 1. We went further by taking 10% of data1 which makes the ratio to the original data 100:1, thus creating data2. We then repeated the same process to produce data3, data4, and data5. Table 2(a) shows the statistics for the original data along with the new generated datasets.

4.2 The Balanced Datasets

By using RUS on the negative class, Table 2(b) shows the five new balanced datasets derived from the Table 2(a) datasets. Note that the size of the data decreased rapidly in some cases. For example, when comparing the size of the class percentage of 0.001, we can see that the HEPMASS dataset has dropped from 5,249,929 to a total number of 105 instances.

4.3 Simulated Experiment Design

Figure 2 outlines the creation of the datasets from Table 2 and the implementation of the ML models from Table 3. After collecting datasets shown in Table 1, our experiment consists of three distinct stages:

- Preparing and sampling the data, where both steps in this stage are repeated five times. By the end of this stage, 50 versions of the original datasets are generated.

  1) Imbalance the full original datasets by randomly discarding samples of the positive class, generating five different class ratios. At this step, we simulated the problem of class imbalance.

  2) Balance these imbalanced datasets into 50:50 class ratios using RUS on the negative class.

Table 2: Random undersampled (RUS) datasets class distribution.

Data	#	Negative Class		Positive Class		Ratios	Total	
		%	Instances	%	Instances	Neg:Pos	%	Instances
HEPMASS	α	100	5,249,876	100	5,250,124	1:1	100	10,500,000
	1	⋮	⋮	10	525,012	10:1	54.999	5,774,888
	2			1	52,501	100:1	50.499	5,302,377
	3			0.1	5250	1000:1	50.049	5,255,126
	4			0.01	525	19,999:2	50.004	5,250,401
	5			0.001	53	699,976:7	49.999	5,249,929
HIGGS	α	100	5,170,877	100	5,829,123	8:9	100	11,000,000
	1	⋮	⋮	10	582,912	71:8	52.307	5,753,789
	2			1	58,291	621:7	47.538	5,229,168
	3			0.1	5829	887:1	47.061	5,176,706
	4			0.01	583	35,483:4	47.013	5,171,460
	5			0.001	58	709,661:8	47.009	5,170,935
SUSY	α	100	2,712,173	100	2,287,827	6:5	100	5,000,000
	1	⋮	⋮	10	228,783	83:7	58.819	2,940,956
	2			1	22,878	1067:9	54.701	2,735,051
	3			0.1	2288	2371:2	54.289	2,714,461
	4			0.01	229	59,274:5	54.248	2,712,402
	5			0.001	23	118,548:1	54.244	2,712,196
sentiment140	α	100	800,000	100	800,000	1:1	100	1,600,000
	1	⋮	⋮	10	80,000	10:1	55	880,000
	2			1	8000	100:1	50.5	808,000
	3			0.1	800	1000:1	50.05	800,800
	4			0.01	80	10,000:1	50.005	800,080
	5			0.001	8	100,000:1	50.001	800,008

2(a) Imbalanced datasets: 5 new datasets are derived from the original dataset using RUS, undersampling the positive class while retaining all of the negative class.

Table 2 contd. ...

...Table 2 contd.

Data	#	%	Negative Class	Positive Class	Ratio	Total Instances
HEPMASS	1	10	524,987	524,167	50:50	1,049,154
	2	1	52,498	52,501	⋮	104,915
	3	0.1	5249	5250		10,491
	4	0.01	524	525		1048
	5	0.001	52	53		105
HIGGS	1	10	517,087	582,912	50:50	1,099,999
	2	1	51,708	58,291	⋮	109,999
	3	0.1	5.170	5829		10,999
	4	0.01	517	583		1100
	5	0.001	51	58		109
SUSY	1	10	271,217	228,783	50:50	49,999
	2	1	27,121	22,878	⋮	49,999
	3	0.1	2712	2288		5000
	4	0.01	271	229		500
	5	0.01	27	23		50
sentiment140	1	10	80,000	80,000	50:50	160,000
	2	1	8000	8000	⋮	16,000
	3	0.10	800	800		1600
	4	0.01	80	80		160
	5	0.001	8	8		16

2(b) Balanced datasets: we balanced each one of the datasets from Table 2(a). This is achieved by using RUS on the negative class so the number of samples are balanced with the positive class labels.

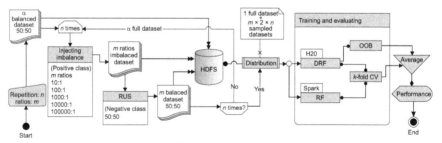

Fig. 2: Simulated experiment design.

- Storing and distributing the data using Apache Hadoop.[4] Hadoop is a popular framework for working with big data that helps to deal with scalability problems by offering distributed storage, the *Hadoop Distributed File System* (HDFS), which is designed to reliably store very large datasets. For more details, please refer to [42].

---

[4] https://hadoop.apache.org/.

- Building and evaluating RF models on the newly created datasets from the first stage along with original datasets, generating a total of 51 datasets that are used to build each model. These models are built using two ML implementation frameworks (Spark and H_2O) with Random Forest with 50 and 100 trees, and the three different AUC methods. Below, we list the modified model configurations, with other parameters kept as the default.
- Maximum depth of each tree in the forest is set to 20.
- The maximum number of bins used for splitting features is set to 32.
- Number of features to consider for splits at each node is square root.
- Criterion used for information gain calculation is Gini index.
- The sub-sampling rate which specifies the size of the dataset used for training each tree in the forest, as a fraction of the size of the original dataset, is set to two thirds.

Table 3: Simulated case study: Area under the ROC curve (AUC) average results.

Framework		H_2O				Spark	
Validation		OOB		5-folds CV			
	Trees	50	100	50	100	50	100
HEPMASS	α	0.948	0.948	**0.966**	0.948	0.945	0.947
	1	0.943	0.945	0.945	**0.946**	0.944	0.945
	2	0.934	0.937	0.937	**0.938**	0.928	0.931
	3	0.897	0.920	0.920	**0.928**	0.809	0.864
	4	0.618	0.691	0.672	**0.787**	0.547	0.583
	5	0.546	0.549	0.520	**0.560**	0.500	0.505
HIGGS	α	0.819	0.823	0.821	**0.824**	0811	0.815
	1	0.811	0.811	0.803	**0.815**	0.808	0.813
	2	0.769	0.769	0.759	**0.778**	0.752	0.766
	3	0.696	0.714	0.713	**0.724**	0.671	0.694
	4	0.600	0.645	0.640	**0.655**	0.523	0.556
	5	**0.555**	0.554	0.524	0.548	0.500	0.499
SUSY	α	0.873	0.874	0.874	**0.875**	0.874	0.874
	1	0.868	0.870	0.870	**0.871**	0.868	0.870
	2	0.860	0.864	0.864	**0.866**	0.846	0.852
	3	0.803	0.826	0.830	**0.842**	0.711	0.753
	4	0.576	0.614	0.598	**0.654**	0.570	0.588
	5	0.525	0.525	**0.547**	0.525	0.500	0.500
sentiment140	α	0.763	0.788	0.756	0.771	0.809	**0.823**
	1	0.762	0.801	0.760	0.774	0.807	**0.823**
	2	0.707	0.744	0.705	0.717	0.736	**0.764**
	3	0.602	0.684	0.637	0.659	0.702	**0.730**
	4	0.613	**0.689**	0.613	0.622	0.612	0.644
	5	0.605	0.761	0.476	0.449	0.624	**0.772**

3(a) Imbalanced datasets results: average AUC results for the generated datasets shown in Table 2(a).

Table 3 contd. ...

...Table 3 contd.

Framework		H$_2$O				Spark	
Validation		OOB		5-folds CV			
	Trees	50	100	50	100	50	100
HEPMASS	1	0.943	0.945	0.945	**0.946**	0.945	0.945
	2	0.936	0.939	0.940	**0.941**	0.939	0.940
	3	0.926	0.933	0.933	**0.935**	0.932	**0.935**
	4	0.916	0.924	0.925	**0.929**	0.920	0.927
	5	0.845	0.845	0.830	0.840	**0.880**	0.873
HIGGS	1	0.802	0.810	0.811	**0.814**	0.809	0.813
	2	0.778	0.810	0.794	**0.814**	0.791	0.798
	3	0.739	0.756	0.753	**0.766**	0.755	0.765
	4	0.679	0.725	0.720	**0.725**	0.693	0.702
	5	0.598	0.690	0.608	**0.643**	0.631	0.604
SUSY	1	0.866	0.870	0.870	**0.872**	0.869	0.871
	2	0.858	0.863	0.864	**0.867**	0.863	0.866
	3	0.836	0.845	0.846	0.849	0.848	**0.853**
	4	0.829	**0.843**	0.829	0.830	0.831	0.841
	5	0.834	0.875	0.838	0.864	**0.916**	**0.916**
sentiment140	1	0.773	0.794	0.806	0.817	0.802	**0.819**
	2	0.756	0.792	0.797	**0.802**	0.784	**0.802**
	3	0.741	**0.764**	0.745	0.762	0.710	0.748
	4	0.599	0.604	0.548	0.615	**0.629**	0.626
	5	0.432	0.386	0.246	0.280	**0.442**	0.325

3(b) Balanced datasets results: average AUC results for the Table 2(b).

A total of 204 datasets are used in this process, from the four big datasets listed in Table 1. We believe that there are benefits in having the original dataset. Many researchers have an imbalance problem then adjust the datasets with sampling, either oversampling or undersampling, to mitigate the effects of class imbalance on model performance. This case study, on the other hand, has the original full balanced datasets to have as comparison with our undersampled dataset results.

4.4 *Results of the Simulated Datasets*

Table 3 presents average AUC results for the iterations and each combination. Figure 3 visualizes the overall AUC slopes for a better understanding of the results. With respect to best performances among all five sampled ratios, the datasets can be categorized into four levels of difficulty. HEPMASS was the easiest to learn among all four, while sentiment140 was the most difficult. With regards to the number of trees in the Random Forest models, as expected, 100 trees performed better than 50.

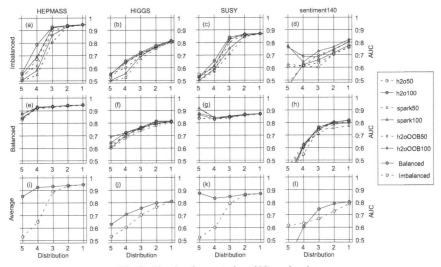

Fig. 3: Simulated case study: AUC results plots.

From Table 3 and Figure 3(a, b, c, and d), we can see that building the models using cross-validation performed better, in general, than using OOB. Note that the AUC values in bold are the best with regards to each generated data in the rows. Upon inspection, we find that the H_2O version of RF, with 100 trees, performed the best on all datasets except sentiment140 where Spark with 100 trees performed better. Regardless, our goal is not an ML library and framework comparison. In terms of the imbalanced datasets, the worst results were at or below 0.1% of positive class membership. We can see from Figure 3(a, b, c, and d) that the sampled positive percentage of 0.1% and 1.0% gave similar performances to the results from 10% and even close to 100% which is the original data. In terms of sentiment140, the AUC is poor at 0.001% of the positive class. The nature of this data is different from the other three datasets and, to be specific, 18 positive class instances is very low especially given the high-dimensionality of the data. On the other hand, Figure 3(e, f, g, and h) shows the balanced class distribution dataset results from Table 3(b) and overall, performance increased. When comparing the same dataset, sentiment140, from graph h with the one from d, we can see that at the smallest class distribution, the AUC results were below 0.5 which means that the built models using 16 samples did not perform well despite the fact that all platforms and learners agreed. Note that the results from H_2O and Spark are not the same due to several factors such as different implementations of the RF algorithm, randomness in the bagging and feature selection, as well as different ways of handling categorical variables.

Figure 3(i, j, k, and l) depicts the results from Table 3. The dashed lines represent the average AUC for the imbalanced datasets for every combination of the five generated ratios. The solid lines represent the average after balancing the datasets. For the first two combinations, 1 and 2, we can see results are fairly similar between the imbalanced and balanced datasets. However, we can see a noticeable increase in performance with the remaining three combinations. This agrees with the results from [13] which concluded that sampling produces an improvement in the overall performance for many classifiers. However, a limitation of the sentiment140 dataset is that the performance of the models was poor and unstable when the positive minority class was at or below 0.01%. Thus, with imbalanced data, it is important to analyze some data characteristics that interact with this issue, aggravating the problem in order to increase the performance.

We are interested in determining whether balancing several class distributions using RUS on big data has an effect on the performance. Table 4 part (a) shows a two-factor ANOVA which includes the different class distributions and whether they are balanced. In these results, based on the p-values and a significance level of 0.05, the p-value for class distribution was 2e-16 which indicates that the levels are associated with different significant strengths. Also, the p-value for the balanced condition was 0.00339, which is also lower than 0.05 indicating a significant difference.

A post hoc test is needed in order to determine which groups differ from each other. The phrase "post hoc" refers to the fact that these tests are conducted without any particular prior comparisons in mind. Table 4(b, and c) presents Tukey's *Honestly Significant Different* (HSD) post hoc tests for balanced and class distributions treatment. In part (b) of Table 4, which tests the balanced criteria with 400 runs in each category, the two groups hold distinct group letters which means there is a significant difference between the balanced and imbalanced class distributions. On the other hand, in part (c) of the Table, the test indicates that there is a degree of performance similarity with some of the RUS class distributions; however, despite the fact that group letters have interactions in most of the cases, mean AUC values of imbalanced class distributions are always preceded by balanced class distributions that share the same positive class percentage. RUS 10:10 has the best performance in our experiment, and class distributions below a positive class percentage of 0.1% lie at the end of the Table. 100:0.001 class distributions have the worst AUC results among all of the ten.

Figure 4 visualizes the AUC ranges and group letters for each class distribution. The figure corresponds to Table 4 part (c). The range of each class distribution is determined by the minimum and maximum AUC values from the Table and the bold dots represent the mean AUC for each class distribution.

Table 4: Simulated case study: Analysis of variance table.

	Df	Sum Sq	Mean Sq	F value	Pr (> F)
Balanced	1	0.139	0.1394	8.635	0.00339
Distribution	8	2.998	0.3748	23.213	< 2e-16
Residuals	790	12.755	0.0161		

4(a) Two-factor ANOVA results.

	AUC	std	r	Min	Max	Group
Yes	0.7266479	0.16219	400	0.2100	0.9548	a
No	0.7002451	0.11479	400	0.4301	0.8850	b

4(b) Tukey's HSD balanced results.

	AUC	std	r	Min	Max	Group
10:10	0.81731	0.04666	80	0.62034	0.87906	a
100:10	0.79110	0.06870	80	0.61003	0.87298	ab
1:1	0.77558	0.09861	80	0.43907	0.88790	ab
100:1	0.74909	0.07968	80	0.55250	0.87439	bc
0.1:0.1	0.69186	0.16996	80	0.21000	0.86097	cd
100:0.1	0.69169	0.09005	80	0.43005	0.85668	cd
0.01:0.01	0.68885	0.17528	80	0.21667	0.92712	cd
0.001:0.001	0.65963	0.20828	80	0.21000	0.95478	de
100:0.01	0.64921	0.09135	80	0.51872	0.88495	de
100:0.001	0.62015	0.13799	80	0.43005	0.88495	e

4(c) Tukey's HSD class distribution results.

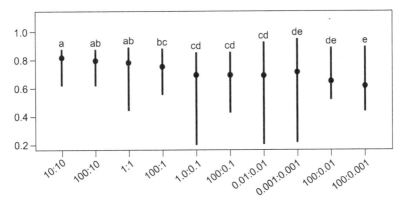

Fig. 4: Simulated case study: Class distribution range and groups.

5. Real-World Implanted Case Study

For the second case study, we use the following three *Public Use File* (PUF) datasets that are related to *Medicare Provider Utilization and Payments* (MPUP) for Medicare fraud detection, with a very limited number of known fraud labels (i.e., a severe class imbalance). In this section, we provide discussions on each of these Medicare datasets, to include data processing and fraud label mapping, as well as our fraud detection results for the original and sampled datasets.

- Medicare Provider Utilization and Payment Data: Physician and Other Supplier (Part B).
- Medicare Provider Utilization and Payment Data: Part D Prescriber (Part D).
- Medicare Provider Utilization and Payment Data: Referring Durable Medical Equipment, Prosthetics, Orthotics and Supplies (DMEPOS).

5.1 Medicare Data Description

1) *Part B*: The Part B dataset provides claims information, within a given year, for each procedure a physician performs. Currently, this dataset is available on the CMS website for the 2012 through 2016 calendar years, with 2016 data being released in 2018 [58]. The years 2017 and 2018 are presently unavailable (for Part B and other Medicare datasets used herein). A unique *National Provider Identifier* (NPI) standard is used to identify physicians, with specific procedures labeled by their *Healthcare Common Procedure Coding System* (HCPCS). The data also includes other claims information which are average payments and charges, the number of procedures performed, and medical specialty. The *Centers for Medicare and Medicaid Services* (CMS) aggregates the data using NPI of the provider, HCPCS code for the procedure, and the place of service. Because physicians may perform the same procedure at different service places and practice under several provider types, for each physician, there are as many records as unique combinations of NPI, Provider Type, HCPCS code, and place of service.

2) *Part D*: The Part D dataset provides information related to the prescription drugs prescribed by physicians and paid for under the Medicare Part D Prescription Drug Program within a given year. Currently, this data is available on the CMS website for the 2013 through 2016 calendar years, with 2016 being released in 2018 [59]. Providers/prescribers are identified using their unique NPI while each drug is listed by its brand and/or generic name along with other information related to the prescription and other general features. Similar to the Part B dataset, we found that physicians practice under

Table 5: Medicare datasets.

Data Name	Neg	Pos	Pos %	Year Range	Features	One-hot
Part B	4,690,862	1508	0.03%	2012–16	35	126
Part D	2,843,498	1153	0.04%	2013–16	34	126
DMEPOS	1,153,265	710	0.06%	2013–16	41	145
Combined	1,015,741	528	0.05%	2013–16	102	137

multiple specialties. There are as many records as unique combinations of NPI, Provider Type, and drug name for each physician. To provide privacy protection for Medicare beneficiaries, an exclusion of any aggregated rows, derived from 10 or fewer claims, has been applied.

3) *DMEPOS*: The *Referring Durable Medical Equipment, Prosthetics, Orthotics and Supplies* (DMEPOS) data includes submitted claims information about medical products for patients based on physicians' orders within a given year. It mainly contains data on utilization, allowed amount and Medicare payment, and submitted charges organized by NPI, HCPCS code, and supplier rental indicator. Currently this data is available on the CMS website for the 2013 through 2016 calendar years (with 2016 being released in 2018) [60]. As previously mentioned for Part B and D, we have found that some physicians place referrals for the same DMEPOS equipment, or HCPCS code, as well as a few physicians that practice under multiple specialties. Therefore, for each physician, there are as many rows as unique combinations of NPI, Provider Type, HCPCS code, and equipment status.

4) *Combined dataset*: The Combined dataset is created after processing Part B, Part D, and the DMEPOS datasets, containing all the attributes from each, along with the fraud labels derived from the LEIE. The combining process involves a join operation on NPI, Provider Type, and year. Due to there not being a gender variable present in the Part D data, we did not include this variable in the join operation conditions and used the gender labels from Part B while removing the gender labels gathered from the DMEPOS dataset after joining. In combining these datasets, we are limited to those physicians who have participated in all three parts of Medicare. Even so, this Combined dataset has a larger and more encompassing base of attributes for applying data mining algorithms to detect fraudulent behavior, as demonstrated in our study.

5) *LEIE*: All three previously listed Medicare datasets (Part B, Part D, and DMEPOS) are aggregated based on the procedure-level and are not labeled with a specific classification problem. In order to generate necessary fraud labels, we integrate information from a list of federally excluded healthcare

Table 6: Mandatory exclusions.

Social Security Act	42 USC	Amendment
1128(a)(1)	1320a-7(a)(1)	Conviction of program-related crimes. Minimum Period: 5 years
1128(a)(2)	1320a-7(a)(2)	Conviction relating to patient abuse or neglect. Minimum Period: 5 years
1128(a)(3)	1320a-7(a)(3)	Felony conviction relating to health care fraud. Minimum Period: 5 years
1128(a)(4)	1320a-7(a)(4)	Felony conviction relating to controlled substance. Minimum Period: 5 years
1128(c)(3)(G)(i)	1320a-7(c)(3)(G)(i)	Conviction of second mandatory exclusion offense. Minimum Period: 10 years
1128(c)(3)(G)(ii)	1320a-7(c)(3)(G)(ii)	Conviction of third or more mandatory exclusion offenses. Permanent Exclusion

providers. The *List of Excluded Individuals and Entities* (LEIE) [61] provides a list of mandatory excluded providers in which the provider is excluded, for a given period of time, from practicing medicine in the United States. This list is issued by the *Office of Inspector General* (OIG) [62] and it is updated monthly. The LEIE only lists provider exclusions without any information regarding which procedures or prescriptions led to being placed on the exclusion list, and is considered a provider- or NPI-level data source. This dataset roughly contains 70,000 records in which only 4900 have a valid NPI, while the remaining are empty. This dataset was mapped with the other Medicare datasets to consider the class-label "exclusion" in which 0 refers to the negative class (non-fraud) and 1 to the positive class (fraud). Table 6 gives the corresponding codes for provider exclusions and the length of each mandatory exclusion. We have determined and assume that any behavior prior to and during a physician's exclusion end date constitutes fraud.

5.2 *Medicare Data Processing*

Each of the aforementioned Medicare datasets require some data preparation prior to building machine learning models. All four Medicare datasets have missing values. The tool we used in the next part does handle missing values; however, its ML library does not handle those missing values automatically; thus, certain transformation were applied on the datasets regarding the missing values. In the datasets, null is used for values that are unknown or missing. Additionally, all standard deviations of NA (i.e., no computed value) were imputed and replaced with 0. Each dataset has several categorical features such as Provider Type, and gender in which those categorical features in the datasets were converted into one-hot encoding. The main reason we followed this method is that applying some ML algorithms such as LR does not consider categorical variables in nature. Thus, indexing these categorical variables may imply a numerical order or value.

With these processed datasets, we map fraud labels from the LEIE. Because the Medicare datasets are annual, we assume that any excluded provider in the LEIE is considered fraudulent for any particular matched year, with a 6-month rounding approach [63]. This is a limitation in the Medicare datasets in both not providing more granular time information and in the LEIE, where no information is given regarding procedures/services associated with each excluded provider. With this assumption, we join providers in the LEIE with each Medicare dataset by NPI and year. Any providers that match between datasets are flagged as a 1 (fraud), otherwise they are flagged as 0 (non-fraud). These are the binary labels used to build models and evaluate fraud detection performance. Note, NPI and year labels are removed after data processing prior to applying any ML approaches.

5.3 *Medicare Experiment Design*

Unlike the first case study, we use three ML learners: LR, RF, and GBT. Additionally, for this experiment, only the Apache Spark machine learning library is used. We only use Spark in this case study and not H_2O, because H_2O does not currently have an included data sampling implementation. The reason for choosing these learners is that they cover several families of algorithms. These learners provide additional insight into the effects of class imbalance on overall machine learning model performance. Moreover, they are generally considered as robust and good learners. In this section, we provide learner configurations, as well as data sampling configurations for the fraud detection experiment. The overall experimental flow is depicted in Figure 5.

1) *Learner Configurations*: With RF and GBT, the number of trees was set to 100 trees. The maximum memory in megabytes (MB) was set to 1024 to speed up model training. CachNodeIds was set to true for speeding up the process of building up the tree. The featureSubsetStrategy parameter was set to one-third based on an initial investigation. Because the categorical features space is converted to one-hot encoding and the feature set is not considered high dimensional, we decided to go with a one-third data subset, because it was found to be better in an initial investigation regarding this particular dataset. Based on initial investigations, we used the Gini index for the information gain calculation. Maximum Bins was set to the maximum number of categorical features, which is 2 in our case, because there is no method to disable it within Apache Spark. All other parameters in RF and GBT were set to their respective default values.

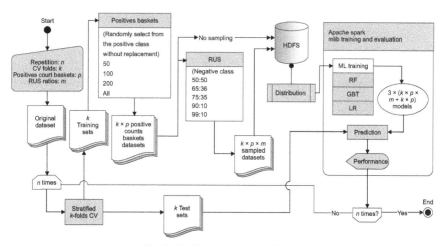

Fig. 5: Medicare experiment design.

The Spark LR max iteration was set to 100. The ElasticNet mixing parameter alpha, in the range [0, 1], was set to 0 indicating an L2 penalty. Apache Spark has impeded standardization which we set to true, determining whether to standardize the training features before fitting the model. Spark LR provides an implementation for tree aggregating which is a specialized implementation of aggregate that iteratively applies the combine function to a subset of partitions. This is done in order to prevent returning all partial results to the driver node where a single pass "reduce" would take place, as with the classic aggregate method. Many of the Spark machine learning algorithms use this tree Aggregate functionality, and show increased model performance.

2) *Data Sampling*: We applied RUS to the following class ratios: 50:50, 65:35, 75:25, 90:10, and 99:1. The ratios are in the form of [negative:positive] classes. The reason we have chosen these ratios is that they cover a good range from balanced datasets to relatively imbalanced datasets. For instance, a 50:50 class ratio for Part B would have 1508 records for the negative class and the same for the positive class. However, a ratio of 99:1 would include 149,293 records of the negative class in the dataset.

Prior to sampling, we generated several extreme positive class counts (or baskets) in which we randomly picked a number of records while discarding the rest. We selected 50, 100, and 200 positive count baskets for each of the Medicare datasets. As an example, a 50:50 class ratio for a basket of 50 positive records would lead to a dataset of only 100 instances, for each of the four datasets. From this process, we studied the rarity of the positive class for which we injected a severe imbalance into the datasets [64], [65]. Note that the label "ALL" includes all available positive class instances. With four datasets, four (baskets) positive counts, three learners, six ratios, 5-fold CV, and 10 repetitions, we built and evaluated 14,400 different models.

5.4 Medicare Fraud Detection Results

As with the previous case study, we also provide Figure 6 which shows the average AUC results for each dataset and number of positive class instances. From these plots, there are noticeable differences in performance across the sampling ratios for each dataset. A post hoc test was applied in order to determine which groups differ significantly from another. Our main goal in investigating this real-world dataset is to determine if these results agree with the conclusions found via the simulated experiments in the first case study. As seen in Table 7, the most important factors are found to be the different baskets of positive classes and the sampling ratios.

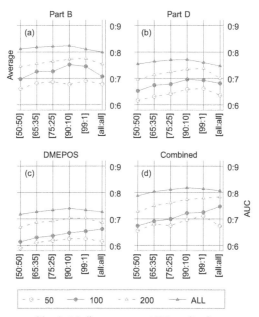

Fig. 6: Medicare average AUC results plots.

Table 7: Medicare 3-factor ANOVA.

	Df	Sum Sq	Mean Sq	F Value	Pr (> F)
isSampled	1	0.13	0.127	26.30	2.95e-07
Rare Baskets	3	32.29	10.765	2228.36	< 2e-16
Ratio	4	1.74	0.435	90.01	< 2e-16
Residuals	14391	69.52	0.005		

Table 8 represents the Tukey's HSD post hoc test results, where boldface values are the highest average AUC scores per factor. It also shows the average AUC, standard deviation, minimum, maximum, and quantiles for each factor. The Part B dataset has the highest AUC value among the four datasets, while DMEPOS has the lowest. Also, as expected, using all of the positive class instances for each dataset yields a higher performance than using any of the generated baskets. Among the three learners, LR performs better on average while RF and GBT perform similar to each other. Additionally, the "isSampled" factor, which refers to the condition that the data was randomly undersampled or kept without sampling, shows that RUS performs better than datasets without sampling. However, some of these non-sampled datasets have very low positive class counts, such as 50, making it extremely difficult for a model to discern distinct positive class patterns. Lastly, the factor "ratio" shows that using a 50:50 class ratio is not always ideal, as it generally depends on other factors such as the data domain, such as Medicare fraud dedication in our case. Under the same Tukey's HSD group, class ratios of RUS 90:10 and 99:1 performed better on average by 0.02 AUC than the other class ratios.

6. Conclusions

The importance of big data is increasing due to the ease of acquiring such data, particularly in fields such as healthcare. Big data is typically defined by a very large amount of information with various complex characteristics. Given that, traditional data mining approaches might not cope with the requirements imposed by big data. In this study, we focus on a major challenge in the Data Mining and Machine Learning communities, namely class imbalance in big data. This problem leads to additional demands on and complexity in the data when training and evaluating machine learning models. We discussed two case studies in which we decreased the size of various big datasets to study the impact of data sampling in favoring the minority (positive) class, which is usually the class of interest.

With the first case study, we deliberately injected and simulated a binary imbalanced classification problem, in which we compared several class ratios and discussed the impact on a RF model's predictive performance. We collected four public balanced big datasets and randomly discarded instances from the positive class, generating five different class ratios. Following this, we employed RUS to balance the negative and positive classes to a 50:50 class ratio on all previous imbalanced datasets. In the second case study, we introduced a real-world case study involving Medicare fraud detection. Five RUS class ratios were created, which include 50:50, 25:75, 35:65, 90:10,

Table 8: Medicare results.

Factor	Level	AUC	std	r	Min	Max	Q25	Q50	Q75	Groups
Dataset	PartB	**0.74462**	0.07864	3600	0.40208	0.93332	0.70071	0.76334	0.80761	a
	Combined	0.73524	0.08130	3600	0.35678	0.93234	0.68846	0.75340	0.79757	b
	PartD	0.69931	0.08003	3600	0.32207	0.91411	0.65397	0.71552	0.75875	c
	DMEPOS	0.66838	0.07708	3600	0.32441	0.87679	0.62523	0.68598	0.72571	d
Rare Baskets	ALL	**0.77753**	0.04089	3600	0.65533	0.87018	0.74412	0.78300	0.81238	a
	200	0.73122	0.05395	3600	0.51942	0.88912	0.69483	0.73437	0.77033	b
	100	0.68781	0.07518	3600	0.42523	0.93332	0.63724	0.69031	0.74261	c
	50	0.65099	0.09798	3600	0.32207	0.93234	0.58415	0.65274	0.72006	d
Learner	LR	**0.72634**	0.08179	4800	0.32207	0.93234	0.67886	0.74159	0.78876	a
	GBT	0.70473	0.08501	4800	0.32441	0.92224	0.65476	0.71907	0.76492	b
	RF	0.70459	0.08589	4800	0.32287	0.93332	0.65250	0.71595	0.76550	b
isSampled	RUS	**0.71321**	0.08431	12000	0.32207	0.93332	0.66307	0.72658	0.77510	a
	NO	0.70524	0.08725	2400	0.32287	0.90951	0.65165	0.71721	0.77313	b
Ratio	[90:10]	**0.72619**	0.08170	2400	0.36455	0.91411	0.68108	0.73889	0.78367	a
	[99:1]	0.72334	0.08040	2400	0.39293	0.93234	0.67544	0.73616	0.78375	a
	[75:25]	0.71511	0.08362	2400	0.32441	0.92581	0.66611	0.72857	0.77613	b
	[65:35]	0.70902	0.08442	2400	0.34365	0.93332	0.65825	0.72192	0.77163	c
	[all:all]	0.70524	0.08725	2400	0.32287	0.90951	0.65165	0.71721	0.77313	c
	[50:50]	0.69241	0.08700	2400	0.32207	0.90078	0.64187	0.70415	0.75585	d

and 99:1. Additionally, we injected more class imbalance into each dataset by creating three baskets (50, 100, 200 positive instances), to assess model performance on a very rare number of positive classes. Besides RF, we included two additional ML models, LR and GBT.

We found, in the first case study, that if the number of minority class labels is too low, such as 100,000:1, then increasing the ratio from 10,000:1 to 1,000:1 can give a good boost in RF performance. Moreover, partially undersampling the majority class, without balancing the data to a 50:50 class ratio, increases model performance. In the second case study, our results agree with the findings from our simulated experiments in case study one. Moreover, we clearly show that a 50:50 balanced class ratio is not always the ideal dataset. In fact, the 99:10 and 99:1 class ratios seem to indicate better performance depending on the total number of available instances.

We suggest that future work should include an investigation on oversampling to inject various degrees of class imbalance. However, ROS might inject redundant information, and thus other synthetic oversampling methods should be explored. Additionally, we will consider additional performance metrics to evaluate the impact of class imbalance.

Acknowledgements

We would like to thank the reviewers in the Data Mining and Machine Learning Laboratory at Florida Atlantic University. Additionally, we acknowledge partial support by the NSF (CNS-1427536). Opinions, findings, conclusions, or recommendations in this paper are the authors' and do not reflect the views of the NSF.

References

[1] Frank J. Ohlhorst. 2012. Big Data Analytics: Turning Big Data into Big Money. John Wiley & Sons.
[2] Viktor Mayer-Schönberger and Kenneth Cukier. 2013. Big data: A revolution that will transform how we live, work, and think. Houghton Mifflin Harcourt.
[3] James Manyika, Michael Chui, Brad Brown, Jacques Bughin, Richard Dobbs, Charles Roxburgh and Angela H. Byers. 2011. Big data: The next frontier for innovation, competition, and productivity.
[4] Andrew McAfee, Erik Brynjolfsson, Thomas H. Davenport et al. 2012. Big data: The management revolution. Harvard Business Review 90(10): 60–68.
[5] Senthilkumar, S.A., Bharatendara K. Rai, Amruta A. Meshram, Angappa Gunasekaran and S. Chandrakumarmangalam. 2018. Big data in healthcare management: A review of literature. American Journal of Theoretical and Applied Business 4(2): 57–69.
[6] Ian H. Witten, Eibe Frank, Mark A. Hall and Christopher J. Pal. 2016. Data mining: Practical machine learning tools and techniques. Morgan Kaufmann.

[7] Julian D. Olden, Joshua J. Lawler and N. LeRoy Poff. 2008. Machine learning methods without tears: a primer for ecologists. The Quarterly Review of Biology 83(2): 171–193.

[8] Jorge Galindo and Pablo Tamayo. 2000. Credit risk assessment using statistical and machine learning: basic methodology and risk modeling applications. Computational Economics 15(1): 107–143.

[9] Victoria López, Sara del Río, José Manuel Benítez and Francisco Herrera. 2015. Cost-sensitive linguistic fuzzy rule based classification systems under the mapreduce framework for imbalanced big data. Fuzzy Sets and Systems 258: 5–38.

[10] Taghi M. Khoshgoftaar, Chris Seiffert, Jason Van Hulse, Amri Napolitano and Andres Folleco. 2007. Learning with limited minority class data. pp. 348–353. *In*: Machine Learning and Applications. ICMLA 2007. Sixth International Conference on, IEEE.

[11] Joffrey L. Leevy, Taghi M. Khoshgoftaar, Richard A. Bauder and Naeem Seliya. 2018. A survey on addressing high-class imbalance in big data. Journal of Big Data 5(1): 42.

[12] Randall Wald, Taghi M. Khoshgoftaar, Alireza Fazelpour and David J. Dittman. 2013. Hidden dependencies between class imbalance and difficulty of learning for bioinformatics datasets. pp. 232–238. *In*: IEEE 14th International Conference on Information Reuse & Integration (IRI), IEEE.

[13] Haibo He and Edwardo A. Garcia. 2009. Learning from imbalanced data. IEEE Transactions on Knowledge and Data Engineering 21(9): 1263–1284.

[14] Jason Van Hulse, Taghi M. Khoshgoftaar and Amri Napolitano. 2007. Experimental perspectives on learning from imbalanced data. pp. 935–942. *In*: Proceedings of the 24th International Conference on Machine Learning, ACM.

[15] Chris Seiffert, Taghi M. Khoshgoftaar and Jason Van Hulse. 2009. Hybrid sampling for imbalanced data. Integrated Computer-Aided Engineering 16(3): 193–210.

[16] Gustavo E.A.P.A. Batista, Ronaldo C. Prati and Maria Carolina Monard. 2004. A study of the behavior of several methods for balancing machine learning training data. ACM SIGKDD Explorations Newsletter 6(1): 20–29.

[17] Nitesh V. Chawla, Kevin W. Bowyer, Lawrence O. Hall and W. Philip Kegelmeyer. 2002. Smote: synthetic minority over-sampling technique. Journal of Artificial Intelligence Research 16: 321–357.

[18] Chao Chen, Andy Liaw and Leo Breiman. 2004. Using Random Forest to Learn Imbalanced Data. University of California, Berkeley 110.

[19] Miroslav Kubat, Stan Matwin et al. 1997. Addressing the curse of imbalanced training sets: one-sided selection. pp. 179–186. *In*: ICML, Vol. 97, Nashville, USA.

[20] Richard A. Bauder, Taghi M. Khoshgoftaar and Tawfiq Hasanin. 2018. Data sampling approaches with severely imbalanced big data for medicare fraud detection. pp. 137–142. *In*: IEEE 30th International Conference on Tools with Artificial Intelligence (ICTAI), IEEE.

[21] Nitesh V. Chawla. 2009. Data mining for imbalanced datasets: An overview. pp. 875–886. *In*: Data Mining and Knowledge Discovery Handbook, Springer.

[22] Pablo D. Gutiérrez, Miguel Lastra, José M. Benítez and Francisco Herrera. 2017. Smote-gpu: Big data preprocessing on commodity hardware for imbalanced classification. Progress in Artificial Intelligence, 1–8.

[23] Philip C.L. Chen and Chun-Yang Zhang. 2014. Data-intensive applications, challenges, techniques and technologies: A survey on big data. Information Sciences 275: 314–347.

[24] Victoria López, Alberto Fernández, María José del Jesus and Francisco Herrera. 2013. A hierarchical genetic fuzzy system based on genetic programming for addressing classification with highly imbalanced and borderline data-sets. Knowledge-Based Systems 38: 85–104.

[25] Charles Elkan. 2001. The foundations of cost-sensitive learning. pp. 973–978. *In*: International Joint Conference on Artificial Intelligence, Vol. 17. Lawrence Erlbaum Associates Ltd.

[26] Bianca Zadrozny, John Langford and Naoki Abe. 2003. Cost-sensitive learning by cost-proportionate example weighting. pp. 435–442. *In*: Data Mining. ICDM 2003. Third IEEE International Conference on, IEEE.

[27] Reshma C. Bhagat and Sachin S. Patil. 2015. Enhanced smote algorithm for classification of imbalanced big-data using random forest. pp. 403–408. *In*: Advance Computing Conference (IACC), IEEE International, IEEE.

[28] Isaac Triguero, Sara del Río, Victoria López, Jaume Bacardit, José M. Benítez and Francisco Herrera. 2015. Rosefw-rf: the winner algorithm for the ecbdl'14 big data competition: an extremely imbalanced big data bioinformatics problem. Knowledge-Based Systems 87: 69–79.

[29] Isaac Triguero, M. Galar, H. Bustince and Francisco Herrera. 2017. A first attempt on global evolutionary undersampling for imbalanced big data. pp. 2054–2061. *In*: Evolutionary Computation (CEC), IEEE Congress on, IEEE.

[30] Matei Zaharia, Mosharaf Chowdhury, Michael J. Franklin, Scott Shenker and Ion Stoica. 2010. Spark: cluster computing with working sets. Hot Cloud 10: 10–10.

[31] Larry J. Eshelman. 2014. The CHC adaptive search algorithm: How to have safe search when engaging. Foundations of Genetic Algorithms 1991 (FOGA 1) 1: 265.

[32] Sara Del Río, Victoria López, José Manuel Benítez and Francisco Herrera. 2014. On the use of mapreduce for imbalanced big data using random forest. Information Sciences 285: 112–137.

[33] Chris Drummond, Robert C. Holte et al. 2003. C4.5, class imbalance, and cost sensitivity: Why under-sampling beats over-sampling. *In*: Workshop on Learning from Imbalanced Datasets II, Vol. 11. Citeseer Washington DC.

[34] Ross J. Quinlan. 1986. Induction of decision trees. Machine Learning 1(1): 81–106.

[35] Matthew Herland, Taghi M. Khoshgoftaar and Richard A. Bauder. 2018. Big data fraud detection using multiple medicare data sources. Journal of Big Data 5(1): 29.

[36] Richard A. Bauder and Taghi M. Khoshgoftaar. 2018. The detection of medicare fraud using machine learning methods with excluded provider labels. pp. 404–409. *In*: FLAIRS Conference.

[37] Leo Breiman. 2001. Random forests. Machine Learning 45(1): 5–32.

[38] Mikel Galar, Alberto Fernandez, Edurne Barrenechea, Humberto Bustince and Francisco Herrera. 2012. A review on ensembles for the class imbalance problem: bagging-, boosting-, and hybrid-based approaches. IEEE Transactions on Systems, Man, and Cybernetics, Part C (Applications and Reviews) 42(4): 463–484.

[39] Taghi M. Khoshgoftaar, Moiz Golawala and Jason Van Hulse. 2007. An empirical study of learning from imbalanced data using random forest. pp. 310–317. *In*: Tools with Artificial Intelligence. ICTAI 2007. 19th IEEE International Conference on, Vol. 2, IEEE.

[40] Manuel Fernández-Delgado, Eva Cernadas, Senén Barro and Dinani Amorim. 2014. Do we need hundreds of classifiers to solve real world classification problems. J. Mach. Learn. Res. 15(1): 3133–3181.

[41] Xiangrui Meng, Joseph Bradley, B. Yuvaz, Evan Sparks, Shivaram Venkataraman, Davies Liu, Jeremy Freeman, D. Tsai, Manish Amde, Sean Owen et al. 2016. Mllib: Machine learning in apache spark. JMLR 17(34): 1–7.

[42] Sara Landset, Taghi M. Khoshgoftaar, Aaron N. Richter and Tawfiq Hasanin. 2015. A survey of open source tools for machine learning with big data in the hadoop ecosystem. Journal of Big Data 2(1): 24.

[43] Ron Kohavi et al. 1995. A study of cross-validation and bootstrap for accuracy estimation and model selection. pp. 1137–1145. *In*: Ijcai, Vol. 14. Montreal, Canada.

[44] Jin Huang and Charles X. Ling. 2005. Using AUC and accuracy in evaluating learning algorithms. Knowledge and Data Engineering, IEEE Transactions on 17(3): 299–310.

[45] Marina Sokolova and Guy Lapalme. 2009. A systematic analysis of performance measures for classification tasks. Information Processing & Management 45(4): 427–437.

[46] Gudmund R. Iversen and Helmut Norpoth. 1987. Analysis of variance. Number 1. Sage.

[47] John W. Tukey. 1949. Comparing individual means in the analysis of variance. Biometrics, 99–114.

[48] Jason Van Hulse, Taghi M. Khoshgoftaar and Amri Napolitano. 2009. An empirical comparison of repetitive undersampling techniques. pp. 29–34. *In*: Information Reuse & Integration. IRI'09. IEEE International Conference on, IEEE.

[49] Lichman, M. 2013. UCI machine learning repository.

[50] Yanmin Sun, Andrew K.C. Wong and Mohamed S. Kamel. 2009. Classification of imbalanced data: A review. International Journal of Pattern Recognition and Artificial Intelligence 23(04): 687–719.

[51] Pierre Baldi, Peter Sadowski and Daniel Whiteson. 2014. Searching for exotic particles in high-energy physics with deep learning. arXiv preprint arXiv: 1402.4735.

[52] Pierre Baldi, Kyle Cranmer, Taylor Faucett, Peter Sadowski and Daniel Whiteson. 2016. Parameterized machine learning for high-energy physics. arXiv preprint arXiv: 1601.07913.

[53] Alec Go, Richa Bhayani and Lei Huang. 2009. Twitter sentiment classification using distant supervision. CS224N Project Report, Stanford 1(2009): 12.

[54] Tomas Mikolov, Kai Chen, Greg Corrado and Jeffrey Dean. 2013. Efficient estimation of word representations in vector space. arXiv preprint arXiv: 1301.3781.

[55] Shengzhe Li, Changlong Jin, Hakil Kim and Stephen Elliott. 2011. Assessing the difficulty level of fingerprint datasets based on relative quality measures. pp. 1–5. *In*: Hand-Based Biometrics (ICHB), International Conference on, IEEE.

[56] Jerzy Stefanowski. 2016. Dealing with data difficulty factors while learning from imbalanced data. pp. 333–363. *In*: Challenges in Computational Statistics and Data Mining, Springer.

[57] David J. Dittman, Taghi Khoshgoftaar, Randall Wald and Amri Napolitano. 2013. Gene selection stability's dependence on dataset difficulty. pp. 341–348. *In*: Information Reuse and Integration (IRI), IEEE 14th International Conference on, IEEE.

[58] Cms: Medicare provider utilization and payment data. physician and other supplier. https://www.cms.gov/Research-Statistics-Data-and-Systems/Statistics-Trends-and-Reports/Medicare-Provider-Charge-Data/Physician-and-Other-Supplier. html. Accessed: 2018-11-20.

[59] Cms: Medicare provider utilization and payment data: Part d prescriber. https://www.cms.gov/Research-Statistics-Data-and-Systems/Statistics-Trends-and-Reports/Medicare-Provider-Charge-Data/Part-D-Prescriber.html. Accessed: 2018-11-20.

[60] Cms: Medicare provider utilization and payment data. referring durable medical equipment, prosthetics, orthotics and supplies. https://www.cms.gov/Research-Statistics-Data-and-Systems/Statistics-Trends-and-Reports/Medicare-Provider-Charge-Data/DME. html. Accessed: 2018-11-20.

[61] Oleie: Office of inspector general leie downloadable databases. https://oig.hhs.gov/exclusions/exclusions_list.asp. Accessed: 2018-11-20.

[62] Office of inspector general exclusion authorities us department of health and human services. https://oig.hhs.gov/. Accessed: 2018-11-20.

[63] Richard A. Bauder and Taghi M. Khoshgoftaar. 2018. A survey of medicare data processing and integration for fraud detection. pp. 9–14. *In*: Information Reuse and Integration (IRI), IEEE 19th International Conference on, IEEE.

[64] Bauder, R.A., T.M. Khoshgoftaar and T. Hasanin. 2018. An empirical study on class rarity in big data. pp. 785–790. *In*: 17th IEEE International Conference on Machine Learning and Applications (ICMLA), Dec 2018.

[65] Chris Seiffert, Taghi M. Khoshgoftaar, Jason Van Hulse and Amri Napolitano. 2007. Mining data with rare events: a case study. pp. 132–139. *In*: Tools with Artificial Intelligence, ICTAI 2007. 19th IEEE International Conference on, Vol. 2, IEEE.

Chapter **2**

How to Optimally Combine Univariate and Multivariate Feature Selection with Data Sampling for Classifying Noisy, High Dimensional and Class Imbalanced DNA Microarray Data#

Ahmad Abu Shanab and *Taghi M Khoshgoftaar**

1. Introduction

The emergence of DNA microarray chips has allowed scientists to measure the expression levels of thousands of genes simultaneously. Practitioners have used machine learning techniques to analyze the data from microarray experiments (gene expression data) and make diagnostic and/or prognostic decisions. However, the extremely large number of genes makes traditional machine learning techniques inefficient and ineffective. With a large number of features, these techniques become computationally expensive and time consuming. Additionally, it is expected that many of these features are irrelevant (having little or no correlation with the class) or redundant

This paper is a revised and expanded version of a paper entitled 'Is Gene Selection Enough for Imbalanced Bioinformatics Data?' [1] presented at the '19th IEEE International Conference on Information Reuse and Integration', Salt Lake City, Utah, USA, 7–9 July 2018.

College of Engineering & Computer Science, Florida Atlantic University, Boca Raton, Florida.
Email: aabusha@fau.edu
* Corresponding author: khoshgof@fau.edu

(containing information already represented in other features) in relation to the question at hand, subsequently leading to suboptimal results (reduced performance and interpretability of predictive models). Feature selection is the main technique used to cope with high dimensionality, which consists of finding a minimum subset of features that are highly correlated with the class attribute. Benefits of feature selection include: enhanced generalization capability of models, improved model interpretability, and accelerated learning time. For these reasons, feature selection has become the cornerstone of data mining in bioinformatics.

Class imbalance is another common challenge in bioinformatics, which occurs when one class, usually the class of interest (i.e., positive class), has fewer instances than the other class(es). This unequal class distribution often results in a large number of false negatives (misclassifications from the positive class), because traditional classifiers were designed with the goal of maximizing overall classification accuracy without properly balancing the weight of each class. Data sampling is the most popular technique to alleviate the problem of class imbalance, which attempts to reduce the severity of imbalance within the data by adding or removing instances. Despite the prevalence of class imbalance among gene expression datasets, most previous studies have ignored the subject entirely or provided shallow treatments. This study shows the importance of taking into account class imbalance when analyzing bioinformatics datasets.

Noise is another challenge exhibited by many real-world datasets, which refers to missing or incorrect values for one or more properties that describe an instance in a dataset. There are two types of data noise: attribute noise and class noise. Attribute noise occurs when values in the independent attributes are incorrect (for example, gene expression levels not recorded correctly), while class noise refers to incorrect values in the dependent attribute (for example, cancerous instances labeled as non-cancerous). Unfortunately, noise has a detrimental impact on classification algorithms as well as feature selection techniques, confusing data mining techniques and subsequently leading to suboptimal results (e.g., worsened classification performance, unstable feature selection). Considering the adverse impact of data noise, there is clearly a need to study its impact on data mining techniques. Thus, all empirical investigations presented in this study were performed on data which was first determined to be free of noise and then had artificial class noise added in a controlled fashion. This way, the results can be used to simulate real-world scenarios.

In this study, we determine whether the order in which feature selection and data sampling are applied is important or not by comparing three approaches developed for classification problems on datasets that exhibit both high dimensionality and class imbalance simultaneously [2]. In the first approach, data sampling takes place before feature selection with the training data being built using the selected features and the original data (DS-FS-UnSam). In the second approach, data sampling also takes place first, but then feature selection is performed; however, the training data is built using the selected features and the sampled data (DS-FS-Sam). In the third approach, feature selection is performed first followed by data sampling, with the training data being built using the selected features and the sampled data (FS-DS). Additionally, we investigate the importance of taking into account the problem of class imbalance on bioinformatics datasets by comparing the classification performance of two approaches. In the first approach feature selection (FS) is performed alone (i.e., no data sampling), and then a classifier is built using the selected features. Alternatively, in the second approach (FS-DS), we apply data sampling after performing feature selection, and then a classifier is built using the selected features and the sampled data. All datasets investigated in the study exhibit high dimensionality. Thus, all of the investigated approaches employ feature selection to cope with the high dimensionality challenge.

To compare the aforementioned approaches, we utilize three feature ranking techniques (with three choices of feature subset size for each), one form of filter-based subset evaluation, and wrapper subset selection, as well as a commonly used data sampling technique (Random Undersampling (RUS)). We perform experiments using ten gene expression datasets that were first determined to be relatively free of noise. We then artificially injected noise, creating three levels of data quality (High-Quality, Average-Quality, and Low-Quality), and we build our final models using six different classification algorithms.

The experimental results demonstrate that FS-DS is the best performing approach for all combinations of learners and data quality levels with one insignificant exception. Additionally, FS-DS was most frequently the top performing approach and was never the worst when considering noisy datasets (Average-Quality and Low-Quality datasets). This is a significant finding demonstrating that FS-DS is robust and noise tolerant, which is a desired quality, especially in bioinformatics. On the other hand, DS-FS-Sam was the worst performing approach, on average, regardless of the data quality level. All of these results were confirmed through ANalysis Of VAriance (ANOVA) and Tukey's Honestly Significant Difference (HSD) tests [6]. Finally, our results show that data sampling (in conjunction with feature selection) helped improve the classification performance even more compared to feature selection alone. Based on these findings, we recommend using feature selection followed by data sampling when dealing with datasets that exhibit both high dimensionality and class imbalance simultaneously.

The remainder of this paper will be organized as follows: Section 2 presents related works on the topics of high dimensionality, class imbalance, and data noise. Section 3 outlines the methods used in this work, the three approaches, the sampling technique, the feature selection techniques, the quality of data, the noise injection mechanism, the datasets, the classifiers, and the performance evaluation. In Section 4, we present our results. Finally, Section 5 concludes our paper and discusses the potential for future work.

2. Related Work

Having a large number of features in a dataset is commonly known as high dimensionality. This overabundance of features makes the process of analyzing such datasets more challenging (requiring extensive computation and degrading the predictive performance of inductive models). Feature selection is the most popular process for handling high-dimensional data, which tries to choose the best features for performing classification and eliminate redundant and useless features. There are a number of advantages when those redundant and irrelevant features are removed, including: enhanced generalization capability of models, improved model interpretability, and a faster learning process.

Feature selection techniques can generally be grouped into two broad categories based on the number of features considered together: univariate techniques and multivariate techniques. Univariate techniques evaluate each feature individually using different statistical measures (filter-based feature ranking), while multivariate techniques evaluate whole subsets at a time either using statistical measures (filter-based subset selection) or using a classifier (wrapper-based feature selection). A broad survey of feature selection is presented by Guyon and Elisseeff [23]. In 2013, our research group conducted a comprehensive study [16] to investigate the effectiveness of 25 different feature ranking techniques and 6 classification algorithms when predicting patient response to a drug treatment. The results showed that the Random Forest classifier is the best performing classifier regardless of the feature selection being used, and it improved classification performance as feature subset size increased.

In the context of subset-based feature selection, Khoshgoftaar et al. [30] investigated the problem of subset-based selection stability (robustness of outputs in the face of perturbation), including the importance of stability as well as various stability measures. The authors investigated the previous studies on stability analysis of feature subset selection techniques within the domain of bioinformatics and have identified the shortcomings of these works to explore possible opportunities for future work. Wald et al. [45] investigated the stability of two filter-based subset selection techniques (Consistency feature subset evaluator and Correlation-Based Feature Selection). They found that Consistency has the greatest stability overall, while Correlation-Based Feature Selection shows moderate stability.

Wrappers received little attention because they can be very computationally expensive and can result in an overfitted inductive model. Inza et al. [27] compared filter-based feature ranking and wrapper-based subset selection. The authors used six feature ranking techniques along with four choices of learner on two bioinformatics datasets. They showed that wrapper feature selection outperforms filter-based ranking; however, it is computationally more expensive. A comparative study on all three forms of feature selection was conducted by Wang et al. [46]. Experiments were conducted using four filter-based rankers, one filter-based subset evaluator, and three classifiers for both wrapper selection and final classification. The authors found that both subset selection approaches (filter-based and wrapper-based) can give good performance while selecting a smaller subset of features.

Class imbalance occurs when positive class instances (that is, those which belong to the most important class) are outnumbered by instances of the other class(es). Many real-world bioinformatics datasets are characterized by class imbalance. Ramaswamy et al. [37] performed feature selection on a dataset where only 16% of the instances are in the class of interest. Shipp et al. [39] classified diffuse large B-cell lymphoma from follicular lymphoma using a dataset with a 25% class imbalance. Iizuka et al. [26] constructed a predictive system using a training dataset of 33 patients, 36% of them belonging to the positive class.

Traditional classifiers applied to class-imbalanced datasets often result in suboptimal classification performance [44]. Data sampling is the most popular technique for handling class imbalanced data [32], where the dataset is transformed into a more balanced one by adding or removing instances. A comprehensive study on different sampling techniques was performed by Kotsiantis [32], Guo [22], and Van Hulse [42], including both oversampling and undersampling techniques (which add instances to the minority class and remove instances from the majority class, respectively), and both random and directed forms of sampling.

Relatively little work focused on both challenges (high dimensionality and class imbalance) together, particularly in the bioinformatics domain. Blagus and Lusa [8] employed three sampling techniques (oversampling, downsizing, and multiple downsizing) as well as variable selection on class imbalanced data. Experiments were conducted using a series of k-NN classifiers along with two linear discriminant classifiers, Random Forest, Support Vector Machine (SVM), CART, a Logistic Regression (LR) based classifier, and prediction analysis of microarrays. The results show that only the k-NN classifiers benefitted from oversampling. The authors considered only one possible order of feature selection and data sampling (named DS-FS-Sam in this work).

In a more recent study, Blagus and Lusa [9] performed a study using data sampling on high-dimensional data. They used two data sampling techniques, RUS and SMOTE, on high-dimensional class-imbalanced breast cancer gene expression datasets and a series of classifiers. They showed that only the k-NN classifiers seem to benefit substantially from SMOTE and a number of the other classifiers seem to prefer RUS. Some of the datasets used in this study were not particularly imbalanced, with the minority class being as high as 45% of the instances. In these cases, data sampling will have little effect as the classes are fairly balanced to begin with. Al-Shahib et al. [4] used undersampling as well as a wrapper-based feature selection to build classifiers to predict protein function from amino acid sequence features. Classifiers were built on the "one versus all" model, with each classifier deciding if instances are in a given class or not. They showed that the classification performance can be improved by combining data sampling and feature selection along with the SVM classifier and that applying the data sampling to improve the class ratio to 50:50 (with or without feature selection) to that same classifier was significantly better than any of the other combinations with few exceptions. This study only considers one possible order of feature selection and sampling, without examining the importance of this order.

Another challenge encountered when analyzing real-world data is noise, which refers to incorrect or missing values in datasets. All kinds of noise can lead to suboptimal classification performance, and class noise has a more harmful effect on classification problems than attribute noise [48]. A comprehensive survey on the sources, challenges, and solutions to address class noise can be found in the work of Frénay and Verleysen [20]. They concluded that many open research questions related to class noise and many avenues remain to be explored. Unfortunately, many data mining techniques are sensitive to data noise. Thus, low quality data can result in suboptimal predictive classification performance and can also impact the effectiveness of feature selection. Therefore, it is important to understand how low quality data can impact data mining techniques (feature selection techniques and classification models). Thus, all empirical investigations presented in this study were performed on data which was first determined to be free of noise and then had artificial class noise added in a controlled fashion. This way, the results can be used to simulate real-world scenarios.

The primary contributions of this paper are as follows: (1) compare three approaches to combining feature selection and data sampling to determine whether the order in which they are applied is important or not, where no previous work systematically investigated the importance of the order for combining feature selection and data sampling in the context of data quality;

(2) investigate the importance of alleviating class imbalance for classification problems on bioinformatics datasets, which have been ignored almost entirely in most previous studies; (3) simulate real-world scenarios by injecting class noise into ten real-world gene-expression datasets (after having been determined to be relatively free of noise) creating three data quality tiers (High-Quality, Average-Quality, and Low-Quality), and (4) examine three major forms of feature selection techniques (filter-based feature ranking, filter-based subset selection, and wrapper subset selection).

3. Methods

This section outlines our experimental methods. Section 3.1 presents the evaluation approaches. Section 3.2 discusses the sampling technique. Section 3.3 presents the 11 feature selection techniques. Section 3.4 describes our measurement for data quality. Section 3.5 describes the datasets used in the work. Section 3.6 outlines our noise injection process. Section 3.7 introduces the learners used to create our classification models. Lastly, Section 3.8 presents the cross-validation process and discusses the performance metric used in this work.

3.1 Investigated Approaches

1) *Approaches for combining feature selection and data sampling*: Feature selection and data sampling have become necessary steps when analyzing high dimensional class imbalanced bioinformatics datasets. Although, these two techniques have received tremendous attention, most works have utilized them separately. However, applying them in conjunction to improve the classification performance has not been thoroughly explored.

We investigated three approaches that are used to deal with both high dimensionality and class imbalance. All approaches combine feature selection and data sampling; the difference between one approach and another is the order (whether sampling takes place before or after feature selection) and the dataset (unsampled or sampled) used for classification. We excluded two other approaches, where only one technique (feature selection or data sampling) is used alone, because all datasets investigated in this paper are imbalanced and exhibit high dimensionality. Both feature selection and sampling are necessary to help alleviate class imbalance and cope with high dimensionality.

The three approaches are outlined in Figure 1. In the first approach (DS-FS-UnSam), data sampling takes place before feature selection is performed, and then a classifier is built using the selected features and the original (unsampled) data. In the second approach (DS-FS-Sam), data sampling also takes place before feature selection is performed; however, a classifier is built using the selected features and the sampled data. On the other hand, in the third approach (FS-DS), feature selection takes place before data sampling is performed, and then a classifier is built using the selected features and the sampled data.

2) *Approaches to investigate the importance of alleviating the class imbalance*: To investigate the importance of alleviating the class imbalance when analyzing bioinformatics datasets we compare two approaches. These two approaches are outlined in Figure 2. The first approach (FS) consists solely of feature selection, and then a classifier is built using the selected features.

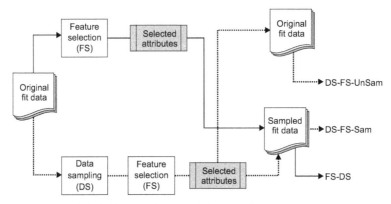

Fig. 1: Feature selection and data sampling approaches.

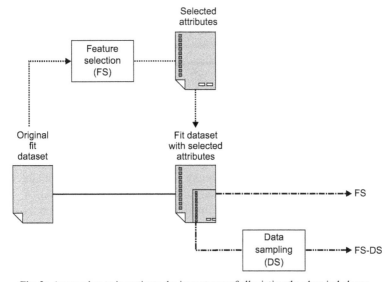

Fig. 2: Approaches to investigate the importance of alleviating the class imbalance.

In the second approach (FS-DS) [2], feature selection takes place before data sampling is performed, and then a classifier is built using the selected features and the sampled data. We selected the FS-DS because our experimentation showed that it is the best approach for utilizing feature selection and data sampling. In summary, the difference between the two approaches is based on whether we employ data sampling or not. This way, the results can be used to determine if data sampling is beneficial in improving the performance for classification models built with bioinformatics datasets.

3.2 Sampling Technique

Data sampling is the process of balancing the class distribution to counter the problem of class imbalance, either by adding (i.e., oversampling) or removing (i.e., undersampling) instances until the desired class ratio is achieved. In this study, we used RUS, which deletes instances randomly from the majority class until the class ratio is balanced at a 50:50 (majority:minority) class ratio. RUS reduces the dataset size, which makes subsequent analysis computationally more efficient compared to oversampling techniques. Additionally, prior research showed its effectiveness [42].

3.3 Feature Selection

In this study we investigated both univariate and multivariate feature selection techniques. In particular, we examined three filter-based feature ranking techniques (with three choices of feature subset size for each), one form of filter-based subset evaluation, and wrapper subset selection.

With the two subset evaluation-based groups (filter-based subset evaluation and wrapper subset selection) a search technique must be used to explore the space of all possible feature subsets in order to reduce the problem from being $O(2^n)$. Based on preliminary experimentation, we chose the Greedy Stepwise approach [12]. This algorithm performs forward selection to build the full feature subset starting from the empty set and stops when none of the new sets outperform the previous best-known set, or when a user-defined maximum number of features (in our study, 100) is reached.

1) *Filter-Based Feature Ranking*: We selected three filter-based feature ranking techniques (i.e., rankers) from three different families: "commonly used" rankers (Chi Squared (CS)), "threshold-based" feature rankers (Area Under the Receiver Operating Characteristic (ROC) Curve), and First Order Statistics-based techniques (Wilcoxon Rank Sum (WRS)). "Threshold-based" feature rankers were proposed and implemented recently by our research group. Readers are referred to the work of Van Hulse et al. [43] for additional information. First Order Statistics-based techniques exhibit the use of first

order statistical measurements such as mean and standard deviation. Thus, in 2012 our research group combined them under this name [29]. Additionally, three choices of feature subset size for each ranker were used (25, 50, and 100). These sizes were proven to be reasonable in a previous study [15]. A brief description of each ranker family is provided below.

Chi Squared (CS) [33] is a statistical test that determines whether there is a statistical relationship between each feature and the class attribute. Chi-squared is found using the following formula:

$$\mathcal{X}^2 = \sum_{i=1}^{I} \sum_{j=1}^{B} \frac{\left[A_{ij} - \dfrac{R_i \times B_j}{N} \right]^2}{\dfrac{R_i \times B_j}{N}}$$

In this equation, I denotes the number of intervals, B the number of classes, N the total number of instances, while R_i is the number of instances in the ith interval, B_j the number of instances in the jth class, and A_{ij} the number of instances in the ith interval and jth class.

Area Under the ROC [36] is a "threshold-based" feature ranker which uses the normalized feature values to classify instances by varying the classification threshold (e.g., instances are considered positive when the feature value is greater than the threshold, otherwise instances are considered negative class examples) to plot the True Positive Rate and the False Positive Rate over all threshold values. The area under the plotted curve determines the quality of the feature. Note that no actual classifier is being built.

Wilcoxon Rank Sum [11] (WRS) is a nonparametric alternative to the standard t-test, in which no assumptions are made about the distribution of the data or population. Instances from both classes are combined and then sorted based on the feature value from smallest to largest, and then each instance will be assigned a rank. The summation of all ranks of the positive instances is computed (i.e., Wilcoxon statistic), then the p-value associated with that Wilcoxon statistic is found from the Wilcoxon rank sum distribution to identify statistically significant features.

2) *Filter-Based Subset Evaluation*: Correlation-Based Feature Selection (CFS) [24] is a commonly used filter-based subset selection technique that is capable of detecting the correlation between features and the class while accounting for the correlation among the features. CFS uses the Pearson correlation coefficient (a measure of the intensity of the linear association between variables). The Pearson correlation coefficient is defined as:

$$M_S = \frac{k\overline{r_{cf}}}{\sqrt{k + k(k-1)\overline{r_{ff}}}}$$

In this equation, M_S is the merit of the current feature subset, k is the number of features, r_{cf} is the mean of the correlations between each feature and the class, and r_{ff} is the mean of the pairwise correlations between every two features.

3) *Wrapper-Based Subset Selection*: Wrapper-based subset selection evaluates feature subsets by applying an induction algorithm and measuring the performance using a classification performance metric. The best performing subset is selected to build the final prediction model, which is usually the same induction algorithm. Although wrappers are computationally expensive, they have the advantage of detecting redundant features. We used Naïve Bayes (NB) (discussed further in Section 3.7) within the wrapper, as it is a simple and effective classification algorithm [18]. To evaluate the classification algorithm with the wrapper we used the Area Under the ROC Curve (AUC) (discussed further in Section 3.8), which previous research showed to be statistically consistent [28].

3.4 Quality of Data

Gene expression datasets are noisy in nature, with difficult to distinguish class boundaries which makes any model-building more difficult. Therefore, there is a clear need to study data mining techniques in the context of data noise. In particular, we create three levels of data quality ("High-Quality," "Average-Quality," and "Low-Quality") to simulate different scenarios and demonstrate the way these approaches would be used in the field. The data quality level is obtained by measuring the classification performance of six commonly used learners: NB, Multilayer Perceptron (MLP), 5-Nearest Neighbor (5NN), SVMs, and two versions of C4.5 decision trees (C4.5 D and C4.5 N) using the AUC performance metric. The average AUC across all learners is used to categorize the dataset(s) according to the following ranges: High-Quality (> 0.8), Average-Quality (≤ 0.8 and > 0.7), and Low-Quality (≤ 0.7). All learners and parameters used are explained in Section 3.7 except for the C4.5 D and C4.5 N learners. C4.5 D is the C4.5 decision tree where the default parameters are used and C4.5 N has pruning turned off and Laplace smoothing turned on. Note that this process is only used to determine the quality level of the raw or noise-injected datasets and does not affect the experiment beyond this measurement.

3.5 Datasets

Ten binary (i.e., each instance is assigned one of two class labels) bioinformatics datasets are considered in this work. All of them are imbalanced, ranging from 10.42% to 35.97% minority instances. Table 1 lists them sorted based on their

Table 1: Dataset characteristics.

Name	# Minority Instances	Total # of Instances	% Minority Instances	# of Attributes	Average AUC
Ovarian Cancer [35]	91	253	35.97%	15155	0.97388
ALL AML Leukemia [40]	25	72	34.72%	7130	0.90908
CNS MAT [13]	30	90	33.33%	7130	0.83551
Prostate MAT [13]	26	89	29.21%	6001	0.90466
MLL Leukemia [40]	20	72	27.78%	12583	0.89615
Lymphoma MAT [17]	19	77	24.68%	7130	0.83659
ALL [40]	79	327	24.16%	12559	0.84748
Lung Clean [3]	23	132	17.42%	12601	0.92351
Lung Cancer [21]	31	181	17.13%	12534	0.93885
Lung Michigan [5]	10	96	10.42%	7130	0.97384

level of class imbalance, as presented in the "% Minority Instances" column. Note that all datasets are high dimensional (number of features ranging between 6,001 and 15,155 features). In addition to the basic properties of each dataset, the table presents the average AUC across the six learners discussed in Section 3.4. All of these datasets have average AUC values greater than 0.8, thus they qualify as High-Quality data according to our measure in Section 3.4.

3.6 Noise Injection

In this work, we created the three levels of data quality ("High-Quality," "Average-Quality," and "Low-Quality") by injecting 24 different class noise patterns into all training datasets. For the noise injection mechanism, the same procedure as reported by Van Hulse et al. [41] is used. Noise is injected in a controlled fashion using two parameters, α (i.e., noise level) and β (i.e., noise distribution). The first parameter controls the total number of noisy instances: $2 \times \alpha \times |P|$ instances will be randomly selected (without replacement) and have their class values switched from positive to negative or from negative to positive (where $|P|$ is the number of minority-class or positive, instances). By tying the number of corrupted instances to the number of minority instances, it can be ensured that they will not overwhelm the minority-class. In this study, we used ($\alpha = 10\%, 20\%, 30\%, 40\%, 50\%$). The second parameter, β determines what fraction of these randomly chosen instances will be selected from the positive class (e.g., $\beta = 0\%$ means that only negative instances are corrupted and $\beta = 100\%$ means that only positive instances are corrupted). This study used ($\beta = 0\%, 25\%, 50\%, 75\%, 100\%$). With five values for α and β, there are 24 different noise injection patterns (because the case with

$\alpha = 50\%$ and $\beta = 100\%$ would convert all positive-class instances into negative-class instances, leaving no counterexamples to learn from).

As mentioned earlier, we injected 24 patterns of class noise into the "raw" (i.e., High-Quality) datasets creating three levels of data quality levels: "High-Quality," "Average-Quality," or "Low-Quality" according to the ranges found in Section 3.4. These categories had 141, 64, and 35 datasets, respectively. We only used the derived datasets in our experiments.

3.7 Classifiers

In this experiment, we used six different classifiers: NB, MLP, 5-NN, SVM, Random Forest with 100 trees (RF100), and LR. We selected these classifiers because they are commonly used in the literature and to include a diverse range of classification algorithms. All classifiers were built using the Weka machine learning software [47], using the default parameters unless noted otherwise. Previous research has shown that the changes described below are appropriate for improving classification models [42].

NB [34] is a simple probabilistic classifier which utilizes Bayes's Theorem of conditional probability and assumes attribute independence. Although this basic assumption is violated in real-world datasets, research has shown that it can be effective and efficient compared to more advanced and sophisticated classifiers. No changes to the default parameters were made in our experiments.

The MLP [7] is a type of neural network that uses backpropagation to classify instances. It contains three layers: an input layer, a hidden layer, and an output layer. In these experiments, the hiddenLayers parameter was set to 3 to build a network with one hidden layer containing three nodes, and the validationSetSize parameter was set to 10 so that the classifier would leave 10% of the instances out to determine when to stop training.

k-nearest neighbors [19], or k-NN, is an example of a case-based learning algorithm, which uses the k closest training samples from a library of all the instances of the training dataset and classifies each new instance to the class most common amongst its k closest neighbors (a k of five was used in this paper, hence the name "5-NN") and the weightByDistance parameter was set to "Weight by $\frac{1}{distance}$".

The SVM [14] is a linear classifier which builds a linear discriminant function using a small number of critical boundary samples from each class while ensuring a maximum possible separation. In Weka, the complexity parameter "c" was changed from 1.0 to 5.0, and buildLogisticModels, which allows proper probability estimates to be obtained, was set to true. In particular, the SVM learner used a linear kernel.

RF100 [10] constructs a large number of unpruned decision trees on randomly bootstrapped data using a randomly-selected subset of features. A new instance is classified by all decision trees and the final classification is induced based on the majority voting. In this study, we changed the numTrees attribute in WEKA to 100 (i.e., 100 trees) and the other parameters were left at the default values.

LR [25] is a statistical regression model for categorical prediction. It predicts the probability of occurrence of an event by fitting data to a logistic curve. The Weka default parameter settings were used for this classifier.

3.8 Performance Evaluation and Cross-Validation

In this study, to avoid the risk of overfitting, we used four runs of five-fold cross-validation [31] to build and test our models. In N-fold cross-validation, the data is randomly split into N mutually exclusive equal-size subsets (folds), and then one of these is held aside as a test (hold-out) fold. The remaining $N - 1$ folds, collectively called the training fold, first had noise injected according to one of the 24 noise patterns, and then models were built on this noisy training fold and classification models were tested on the remaining "clean" fold. A learning algorithm is trained and tested N times. The value $N = 5$ was used in this paper. Once all N folds have been used as the test datasets, the results from all test datasets are integrated into a single performance value for that dataset. Since we are using four runs of five-fold cross-validation, we repeat the feature selection 20 times for each of the derived datasets.

We used the AUC [38] performance metric to evaluate the performance of learners. This performance metric was chosen because it is commonly used in the literature, and due to its invariance to a priori class probability distributions, which makes it suitable when analyzing imbalanced data (note that all datasets in this study exhibit class imbalance). The AUC builds a graph of the True Positive Rate vs. False Positive Rate as the classifier decision threshold is varied, and then uses the area under this graph as the performance across all decision thresholds. Note that while area under the ROC curve is used as both feature ranker (ROC) and as classifier performance metric (AUC), these uses are disconnected from each other.

4. Results

In this work, we compare three approaches for combining feature selection and data sampling (DS- FS-UnSam, DS-FS-Sam, and FS-DS) in Section 4.1. The three approaches differ in the order (whether feature selection takes place before or after data sampling) and the dataset (unsampled or sampled) used to build the training dataset. Additionally, we investigate the importance of alleviating class imbalance by comparing the classification performance of two approaches in Section 4.2. The first approach (FS) does not employ any technique to handle class imbalance and only employs feature selection. On the other hand, the second approach (FS-DS) employs data sampling after performing feature selection. We employed three major types of feature selection (ranker-based techniques, filter-based subset selection, and wrapper-based feature selection). We apply RUS to obtain a balanced class ratio. Additionally, six commonly used classifiers were used to build predictive models. All experiments were performed on 10 bioinformatics datasets which were first determined to be free of noise. We then created three levels of data quality by injecting class noise.

4.1 Importance of the order when feature selection and data sampling are applied

The results are presented in Table 4. Each value represents the average AUC performance across four runs of five-fold cross-validation when applying the given combination of feature selection technique, feature-selection/data-sampling strategy, and classifier to the datasets which match that data quality level. In the "Feature Selection Technique" column, the rankers (CS, ROC, and WRS) are followed by a number, which represents the number of features chosen from that ranked list, and the wrapper-based selection approach which uses the NB learner inside the wrapper is abbreviated as "WrapNB" for space considerations. The table includes six sub-tables: one for each classifier (NB, MLP, 5-NN, SVM, RF100, and LR, respectively). The sub-tables also present the average performance (last row of the sub-tables) of each of the approaches over the 11 feature selection strategies and datasets which match that data quality level for that specific learner. The last row of the table represents the overall average performance of each of the approaches for that specific data quality level. The best and worst choices of approach for each combination of learner and data quality are printed in **bold** and *italics*, respectively.

From the results, we can make the general statement that FS-DS is the best approach to utilize feature selection and data sampling when learning from class imbalanced, high dimensional bioinformatics datasets. The overall average performance shows that FS-DS is the best performing approach across the board (regardless of data quality). When we look at the "Average" row in each sub-table showing the performance across all feature selection strategies, we find that FS-DS is the best performing approach for all combinations of data quality tiers and learners (except High-Quality with LR). The other two approaches did not perform as well: DS-FS-UnSam was in the middle of the performance list on average; for NB, MLP, SVM, and RF100 it was the second best, and was the worst when considering the other learners (5-NN

Table 2: Average AUC values.

Learner	Feature Selection Technique	High Quality			Average Quality			Low Quality		
		DS-FS-UnSam	DS-FS-Sam	FS-DS	DS-FS-UnSam	DS-FS-Sam	FS-DS	DS-FS-UnSam	DS-FS-Sam	FS-DS
NB	CS25	0.956922	0.939667	0.955531	0.862794	0.856155	0.885698	0.739856	0.714112	0.755110
	CS50	0.959370	0.938579	0.952183	0.871306	0.860189	0.888606	0.742820	0.718590	0.748761
	CS100	0.955167	0.934014	0.947035	0.866736	0.855428	0.881821	0.725347	0.712826	0.742586
	ROC25	0.958692	0.946576	0.962710	0.870029	0.867871	0.894704	0.746643	0.722051	0.786812
	ROC50	0.958067	0.943315	0.958141	0.875069	0.872118	**0.894763**	0.751881	0.731324	0.796500
	ROC100	0.952748	0.937655	**0.953093**	0.869952	0.866031	0.888957	0.748590	0.734069	**0.798689**
	WRS25	0.957530	0.945056	0.961441	0.864825	0.862119	0.891515	0.733065	0.711782	0.782320
	WRS50	0.956223	0.941191	0.956522	0.869192	0.866374	0.891053	0.734966	0.720141	0.788267
	WRS100	0.950764	0.934775	0.951179	0.862180	0.858238	0.883945	0.736548	0.723336	0.789965
	CFS	0.937945	0.924392	0.950497	0.811599	0.817122	0.855433	0.669044	0.676202	0.733283
	WrapNB	0.835792	0.830064	0.868756	0.711503	0.707852	0.734158	0.596321	0.593213	0.626226
	Average	0.944836	0.929815	0.947846	0.851993	0.847645	0.874833	0.722660	0.707039	0.760637
MLP	CS25	0.949837	0.940655	0.948744	0.848082	0.834246	0.850272	0.734229	0.707881	0.743435
	CS50	0.953861	0.948902	0.951914	0.852838	0.840810	0.857234	0.740538	0.715005	0.758941
	CS100	0.960967	0.955446	0.958011	0.858286	0.850620	0.866224	0.746738	0.724627	0.759960
	ROC25	0.953674	0.945432	0.958299	0.853856	0.843059	0.861977	0.745979	0.720473	0.767679
	ROC50	0.958871	0.952182	0.959560	0.858792	0.850617	0.865710	0.750513	0.726702	0.769858
	ROC100	0.961491	0.956587	**0.963166**	0.867104	0.855439	**0.871431**	0.750899	0.735905	**0.781452**
	WRS25	0.954065	0.945599	0.957761	0.854462	0.841916	0.861483	0.745195	0.716835	0.769848
	WRS50	0.958447	0.951893	0.959371	0.858027	0.851084	0.866670	0.749886	0.724741	0.770297
	WRS100	0.961224	0.956002	0.962819	0.867409	0.854301	0.870917	0.750887	0.735094	0.778505
	CFS	0.949134	0.946058	0.961668	0.837805	0.833801	0.864029	0.740287	0.712072	0.749076
	WrapNB	0.843205	0.834045	0.880463	0.737496	0.722565	0.763017	0.617948	0.600128	0.635755
	Average	0.947003	0.940450	0.951743	0.847120	0.836561	0.856021	0.735393	0.712320	0.754453

5-NN	CS25	0.953704	0.956645	0.960823	0.847929	0.851715	0.869672	0.710428	0.734122	0.752029
	CS50	0.965272	0.964150	0.966178	0.861466	0.865179	0.880277	0.720440	0.744309	0.768665
	CS100	0.970302	0.968320	0.970497	0.872175	0.872211	0.884749	0.726869	0.746708	0.769050
	ROC25	0.953293	0.955219	0.964155	0.848046	0.856520	0.877677	0.712116	0.736981	0.774097
	ROC50	0.962445	0.962259	0.968287	0.863585	0.872026	0.888707	0.733277	0.750572	0.787125
	ROC100	0.968997	0.966409	**0.972136**	0.874404	0.876468	0.893524	0.741672	0.755795	0.792522
	WRS25	0.953941	0.954880	0.963517	0.847733	0.855716	0.877799	0.705651	0.727978	0.774909
	WRS50	0.961572	0.962071	0.968237	0.864295	0.871577	0.889605	0.726673	0.747163	0.785342
	WRS100	0.968989	0.966003	0.971613	0.876125	0.876875	**0.893934**	0.741471	0.754707	**0.794353**
	CFS	0.958922	0.957368	0.970727	0.845771	0.846970	0.876716	0.710403	0.732188	0.752888
	WrapNB	*0.818756*	*0.832314*	*0.876703*	*0.705934*	*0.717608*	*0.759913*	*0.591255*	*0.600066*	*0.639645*
	Average	0.950087	0.950831	0.960151	0.848828	0.853811	0.874113	0.712529	0.731759	0.764196
SVM	CS25	0.946300	0.929197	0.940573	0.831390	0.823077	0.841703	0.720238	0.705382	0.739134
	CS50	0.939195	0.927435	0.931881	0.824639	0.816886	0.831260	0.719086	0.702378	0.735627
	CS100	0.937636	0.934473	0.933959	0.818697	0.823840	0.829222	0.715291	0.705441	0.735847
	ROC25	0.950437	0.936142	**0.950783**	0.844368	0.834184	**0.858921**	0.730131	0.712864	**0.763731**
	ROC50	0.942167	0.932891	0.940745	0.829304	0.824617	0.844431	0.721214	0.715188	0.754172
	ROC100	0.939026	0.936861	0.942601	0.826409	0.827039	0.843152	0.719614	0.716623	0.759721
	WRS25	0.950232	0.935980	0.950566	0.842009	0.834840	0.858333	0.723477	0.709912	0.763399
	WRS50	0.942757	0.933448	0.940561	0.830201	0.827787	0.846444	0.720455	0.714818	0.756164
	WRS100	0.938757	0.936113	0.942837	0.828340	0.828942	0.844667	0.718393	0.718234	0.758983
	CFS	0.927253	0.929551	0.944706	0.808067	0.811775	0.836028	0.717858	0.713123	0.738035
	WrapNB	*0.817936*	0.836037	0.885095	*0.712097*	0.721493	0.765021	0.582856	0.604834	0.641082
	Average	0.931444	0.925307	0.937245	0.819987	0.817750	0.837661	0.709616	0.702876	0.741618

Table 2 contd....

...Table 2 contd.

Learner	Feature Selection Technique	High Quality			Average Quality			Low Quality		
		DS-FS-UnSam	DS-FS-Sam	FS-DS	DS-FS-UnSam	DS-FS-Sam	FS-DS	DS-FS-UnSam	DS-FS-Sam	FS-DS
RF100	CS25	0.969187	0.955251	0.968309	0.874662	0.855336	0.881142	0.744117	0.713799	0.766278
	CS50	0.976990	0.965916	0.974070	0.894274	0.872044	0.896810	0.768187	0.734437	0.785214
	CS100	**0.981394**	0.971171	0.977398	0.904710	0.881168	0.903292	0.781599	0.752352	0.789484
	ROC25	0.969483	0.958572	0.971536	0.880436	0.864444	0.892014	0.756735	0.733108	0.784975
	ROC50	0.976401	0.966331	0.976160	0.897804	0.879363	0.901876	0.772071	0.749486	0.797153
	ROC100	0.979796	0.971035	0.978867	0.908131	0.886637	**0.908800**	0.788014	0.759363	0.809325
	WRS25	0.969461	0.957710	0.970715	0.879174	0.862433	0.890956	0.748925	0.726660	0.783919
	WRS50	0.976692	0.966586	0.975798	0.896891	0.878118	0.902515	0.771123	0.745215	0.799303
	WRS100	0.980399	0.971477	0.978767	0.906916	0.887170	0.908432	0.786473	0.757678	**0.809843**
	CFS	0.974927	0.963818	0.978352	0.887516	0.862464	0.890485	0.761195	0.731640	0.754519
	WrapNB	*0.832597*	0.838633	0.894984	*0.718018*	0.724336	0.774074	*0.594230*	0.600662	0.628792
	Average	0.963800	0.954486	0.968419	0.879991	0.861942	0.888475	0.754062	0.729319	0.775265
LR	CS25	0.866455	0.868877	0.846261	0.743642	0.751385	0.744122	0.655100	0.669714	0.685226
	CS50	0.840325	0.853312	0.834029	0.719469	0.745230	0.741035	0.643623	0.664317	0.689559
	CS100	0.830151	0.849953	0.834958	0.707030	0.736769	0.735040	0.627445	0.647898	0.681842
	ROC25	0.872804	0.878459	0.874804	0.748057	0.769015	0.766753	0.670578	0.687764	0.714578
	ROC50	0.848052	0.868016	0.862499	0.722330	0.758921	0.755277	0.636923	0.674371	0.704317
	ROC100	0.831921	0.860648	0.856425	0.700410	0.744720	0.745784	0.631843	0.662669	0.697913
	WRS25	0.871226	0.878883	0.876238	0.745140	0.767975	0.766116	0.659711	0.685611	0.717176
	WRS50	0.848156	0.869280	0.862782	0.721510	0.759055	0.757439	0.635926	0.674705	0.705709
	WRS100	0.833725	0.860085	0.858356	0.705496	0.746030	0.747689	0.630497	0.666091	0.699391
	CFS	0.848489	0.900733	**0.902570**	0.715437	0.796670	**0.802972**	0.661949	0.696776	**0.721932**
	WrapNB	0.840103	*0.824609*	0.863791	0.728094	0.707829	0.737880	0.608909	0.597538	0.630244
	Average	0.848400	0.864853	0.860662	0.723389	0.753085	0.753947	0.642275	0.666718	0.695669
Overall	Average	0.930928	*0.927624*	**0.937678**	0.828551	0.828466	**0.847508**	0.712755	0.708339	**0.748640**

and LR), while DS-FS-Sam was the worst performing approach on average. Furthermore, FS-DS showed itself to be particularly noise tolerant by not being at the bottom of the list for Average-Quality and Low-Quality data (higher levels of noise), where it was never the worst performing approach and at worst comes in second place. When considering High-Quality data, FS-DS was the worst performing approach for only 3 of the 66 combinations (SVM learner and the CS ranker with 25 features, and the LR learner with the CS ranker utilizing 25 and 50 features).

Looking closely at these results in terms of the different feature selection techniques, it can be seen that for all subset selection techniques (CFS and Wrapper), the best performing approach was consistently FS-DS regardless of the learner and data quality. The only exception to this is when considering Low-Quality data with the RF100 learner, where FS-DS was the second best. When considering the other category of feature selection (i.e., rankers), we can see that for all but 8 out of 108 combinations of learner and ranker with both Average-Quality and Low-Quality data (Average-Quality with the RF100 learner and CS ranker with 100 features or LR learner and all rankers utilizing 25 and 50 features as well as CS with 100 features), the best approach was FS-DS. This is especially important as these two tiers of data quality represent higher levels of noise. When considering High-Quality data, FS-DS was at the top of the pack for 39 out of 66 combinations. On the other hand, DS-FS-UnSam was the best choice for 20 of 198 combinations, and was the worst for 66 combinations, while DS-FS-Sam was only the best approach for 16 combinations and was at the bottom for 129 of the 198 combinations.

Looking at these results on a per-data quality level basis, we see that FS-DS is particularly robust and is able to improve the classification performance for all learners regardless of the feature selection technique when Low-Quality datasets (AUC less than 0.7 due to noise injection) are used. In particular, FS-DS improved the performance of classifiers enough to result in AUC values greater than 0.7 (which is our metric for Average-Quality) for all combinations of learner and feature selection with few exceptions (e.g., Wrapper regardless of the learner). Additionally, it should be noted that FS-DS was the only approach that was able to improve the classification performance for LR (when combined with ROC25, ROC50, WRS25, WRS50, and CFS feature selection), resulting in AUC values greater than 0.7. FS-DS combined with the RF100 learner helped improve the classification performance on Low-Quality datasets significantly (when combined with ROC100 or WRS100), resulting in AUC values greater than 0.8 (i.e., our metric for High-Quality). Similarly, FS-DS and RF100 improved the performance on Average-Quality datasets, achieving AUC values greater than 0.9.

We performed a set of one-factor ANOVA tests [6] to validate the classification results and found statistically significant outcomes. The ANOVA analysis and subsequent statistical tests were performed within MATLAB®. Since a significance factor of 5% was chosen, the p-value must be less than this value (i.e., 0.05) for the result to be significant.

In this analysis, we considered only one factor: the choice of strategy for combining feature selection and data sampling, with three different levels of this factor (DS-FS-UnSam, DS-FS-Sam, and FS-DS). The tests performed were across all datasets and factors, and for each level of data quality. For the ANOVA tests, the AUC results across all six learners were used as the response variable. The results are presented in Table 3. These results show that the choice of approach for combining feature selection and data sampling is significant across all data quality levels as well as each level of data quality; that is to say, when the data are grouped by the choice of approach, at least two of those groups will have significantly different means.

We wanted to find out which pairs of means are significantly different, and which are not. We conducted a multiple pairwise comparison by using HSD criterion [6]. The significance level for Tukey's HSD test is $\alpha = 0.05$. Figure 3 shows the comparison results of the three choices of approach for combining feature selection and data sampling for all data quality levels, and for each of the different levels of data quality. The results for all datasets, High-Quality datasets only, Average-Quality datasets only, and Low-Quality datasets only are shown in Figures 3a, 3b, 3c, and 3d, respectively. The figures display graphs within each group mean represented by a symbol (∘) and the 95% confidence interval as a line around the symbol. Two means are significantly different if their intervals are disjoint, and are not significantly different if their intervals overlap.

Figure 3 supports our conclusion that the top performing choice of approach for combining feature selection and data sampling is always FS-DS.

Table 3: ANOVA results: Feature-selection/data-sampling strategies across all learners.

Datasets	Source	Sum Sq.	d.f.	Mean Sq.	F	p-value
All Data Quality Levels	FS/DS Strategy	50.4	2	25.1952	1073.41	0
	Error	21727.6	925677	0.0235		
	Total	21777.9	925679			
High Quality	FS/DS Strategy	9.56	2	4.78028	437.57	1.31E-190
	Error	5966.15	546117	0.01092		
	Total	5975.71	546119			
Average Quality	FS/DS Strategy	19.58	2	9.79016	424.17	1.27E-184
	Error	5633.49	244077	0.02308		
	Total	5653.07	244079			
Low Quality	FS/DS Strategy	44.16	2	22.0823	621.47	2.14E-269
	Error	4813.84	135477	0.0355		
	Total	4858.01	135479			

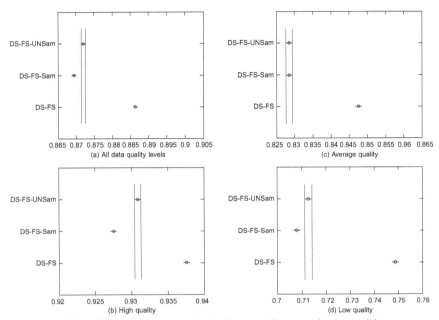

Fig. 3: Tukey HSD results: Feature-selection/data-sampling strategies across all learners.

The difference between the top performing approach and the other approaches (i.e., DS-FS-UnSam and DS-FS-Sam) is statistically significant across all data quality levels. Furthermore, DS-FS-Sam was significantly the worst performing approach across all data quality levels, except when considering Average-Quality datasets, where the difference is statistically insignificant. DS-FS-UnSam, on the other hand, shows average performance on High-Quality and Low-Quality datasets, while being second worst on Average-Quality datasets but not statistically distinguishable.

4.2 *Importance of Alleviating Class Imbalance*

In this section we examine the importance of alleviating class imbalance by comparing two approaches (FS and FS-DS). The results of our experiments can be found in Table 4. Overall, we can make the general statement that in order to improve the performance for classification models built with bioinformatics datasets that exhibit both high dimensionality and class imbalance simultaneously, alleviating class imbalance in conjunction with reducing high dimensionality is the best strategy. The overall average performance shows that FS-DS outperforms FS across the board (regardless of the data quality level). When we look at the "Average" row in each sub-table showing the performance across all feature selection strategies, we find that FS-DS is the best performing approach for all combinations of data

Table 4: Average AUC values.

Learner	Feature Selection Technique	High Quality		Average Quality		Low Quality	
		FS	FS-DS	FS	FS-DS	FS	FS-DS
NB	CS25	0.969516	0.958764	0.895590	0.897738	0.719840	0.730657
	CS50	0.970508	0.956419	0.902930	0.901652	0.719909	0.717874
	CS100	0.970697	0.955836	0.901985	0.900166	0.712186	0.716005
	ROC25	0.979886	0.970850	0.912904	0.917641	0.751071	0.772535
	ROC50	**0.980254**	0.970873	0.911979	**0.917992**	0.754139	0.786815
	ROC100	0.977968	0.971127	0.902382	0.915377	0.742264	**0.788152**
	WRS25	0.979393	0.970763	0.910355	0.915736	0.743097	0.766034
	WRS50	0.979589	0.970059	0.908369	0.914749	0.744373	0.777153
	WRS100	0.977409	0.970455	0.895318	0.909554	0.731238	0.779433
	CFS	0.964765	0.968279	0.856189	0.875105	0.712287	0.727862
	WrapNB	0.873768	0.875268	0.723632	0.740975	0.615231	0.614892
	Average	**0.965796**	0.958063	0.883785	**0.891517**	0.722331	**0.743401**
MLP	CS25	0.957909	0.955020	0.874440	0.872896	0.738998	0.744182
	CS50	0.960888	0.958370	0.870329	0.882694	0.762635	0.766761
	CS100	0.965276	0.970009	0.876488	0.891741	0.766032	0.767741
	ROC25	0.972561	0.972603	0.889551	0.896138	0.772974	0.782343
	ROC50	0.971411	0.974620	0.889696	0.904404	0.773919	0.791021
	ROC100	0.973277	**0.979377**	0.891228	**0.908642**	0.775144	**0.796274**
	WRS25	0.972087	0.972572	0.890120	0.896583	0.776447	0.782432
	WRS50	0.971478	0.974499	0.892507	0.904135	0.779396	0.790007
	WRS100	0.972624	0.978787	0.894527	0.906057	0.777489	0.792431
	CFS	0.972128	0.977827	0.881703	0.895072	0.749556	0.760544
	WrapNB	0.897627	0.881394	0.795654	0.772147	0.652678	0.622512
	Average	0.962479	**0.963189**	0.876931	**0.884592**	0.756842	**0.763295**

5-NN	CS25	0.970207	0.976777	0.856415	0.890983	0.681958	0.744939
	CS50	0.976821	0.981959	0.867209	0.901058	0.691052	0.768582
	CS100	0.983206	**0.986398**	0.881554	0.904065	0.695454	0.771755
	ROC25	0.973206	0.981193	0.871361	0.911186	0.701723	0.786333
	ROC50	0.978648	0.983867	0.885872	0.919467	0.718192	0.794923
	ROC100	0.983968	0.986343	0.896450	0.922232	0.724674	0.802621
	WRS25	0.973461	0.980829	0.871335	0.912627	0.703728	0.788087
	WRS50	0.978743	0.983764	0.886252	0.920598	0.710994	0.793919
	WRS100	0.984148	0.986241	0.897483	**0.922761**	0.724432	**0.805558**
	CFS	0.974902	0.984968	0.856348	0.906968	0.674918	0.762576
	WrapNB	0.871213	0.887178	0.757480	0.779478	0.601233	0.634457
	Average	0.968048	**0.974502**	0.866160	**0.899220**	0.693487	**0.768523**
SVM	CS25	0.949075	0.942009	0.857103	0.859561	0.720630	0.730785
	CS50	0.941267	0.941478	0.838170	0.856167	0.726486	0.742296
	CS100	0.945576	0.953316	0.836320	0.859383	0.731150	0.748844
	ROC25	0.965738	0.963434	0.876210	0.886538	0.752620	0.773483
	ROC50	0.957807	0.959919	0.860315	0.881459	0.751254	0.771793
	ROC100	0.953708	0.965543	0.858416	0.883180	0.749145	**0.778840**
	WRS25	0.965331	0.963588	0.873680	**0.886563**	0.754862	0.771534
	WRS50	0.957439	0.960354	0.860538	0.883548	0.754474	0.773359
	WRS100	0.953602	0.965468	0.858431	0.883186	0.753271	0.778099
	CFS	0.958215	**0.967371**	0.848835	0.868891	0.732260	0.750160
	WrapNB	0.887550	0.886718	0.772574	0.774072	0.621365	0.627824
	Average	0.948664	**0.951745**	0.849145	**0.865686**	0.731593	**0.749729**

Table 4 cond. ....

...Table 4 contd.

Learner	Feature Selection Technique	High Quality		Average Quality		Low Quality	
		FS	FS-DS	FS	FS-DS	FS	FS-DS
RF100	CS25	0.979556	0.976871	0.893113	0.896443	0.739130	0.759041
	CS50	0.984723	0.982120	0.909327	0.914123	0.759805	0.778019
	CS100	0.989225	0.986335	0.922937	0.922435	0.773646	0.783330
	ROC25	0.983257	0.978855	0.908410	0.915990	0.767508	0.785381
	ROC50	0.986803	0.985208	0.920943	0.927072	0.782087	0.794035
	ROC100	**0.989852**	0.987553	0.926962	0.931585	0.793339	0.803700
	WRS25	0.982729	0.977889	0.908789	0.916362	0.764786	0.781745
	WRS50	0.987269	0.984265	0.922320	0.926414	0.783083	0.798303
	WRS100	0.989569	0.987996	0.927036	**0.931910**	0.795667	**0.804954**
	CFS	0.989203	0.985364	0.920940	0.905618	0.750484	0.747825
	WrapNB	0.880631	0.888229	0.781657	0.779135	0.611871	0.609267
	Average	**0.976620**	0.974608	0.903858	**0.906099**	0.756491	**0.767782**
LR	CS25	0.853032	0.865661	0.774650	0.795265	0.657146	0.715298
	CS50	0.829788	0.876675	0.725889	0.789230	0.645076	0.724821
	CS100	0.834142	0.887806	0.717485	0.802440	0.635654	0.729219
	ROC25	0.885725	0.921165	0.788996	0.826819	0.683913	0.754559
	ROC50	0.864656	0.926523	0.749217	0.817484	0.659120	0.755278
	ROC100	0.850500	0.923279	0.740576	0.812882	0.643436	0.752656
	WRS25	0.885544	0.923709	0.787911	0.827983	0.683255	**0.757058**
	WRS50	0.863800	0.925990	0.752706	0.819583	0.658126	0.756025
	WRS100	0.849328	0.923462	0.737418	0.814444	0.638876	0.753414
	CFS	0.867437	**0.952627**	0.730162	**0.842181**	0.654826	0.747270
	WrapNB	0.887879	0.872141	0.760602	0.743522	0.638327	0.621363
	Average	0.861075	**0.909003**	0.751419	**0.808348**	0.654341	**0.733360**
Overall	Average	0.947114	**0.955185**	0.855216	**0.875910**	0.719181	**0.754348**

Table 5: z-test results.

Datasets	z-value	p-value
All Data Quality Levels	−26.2156616	< 0.0001
High Quality	−15.29787687	< 0.0001
Average Quality	−16.95388351	< 0.0001
Low Quality	−21.45216765	< 0.0001

quality tiers and learners (except High-Quality with RF100 and NB). For all but 2 of 18 combinations of learner and data quality level (High Quality with the NB and RF100 learners) the best approach was FS-DS. It is of note that FS-DS consistently outperformed FS (regardless of the data quality level and feature selection technique) when the 5-NN learner is used.

We also performed a set of two-tailed z-tests for each paired comparison to find statistically significant patterns. The tests performed were across all datasets and factors, and for each level of data quality. The z-test method tests the null hypothesis that the population means related to two independent group samples are equal against the alternative hypothesis that the population means are different. p-values are provided for each pair of comparisons in the table. The significance level is set to 0.05; when the p-value is less than 0.05, the two group means are significantly different from one another.

The results are presented in Table 5. These results support our conclusion that the top performing choice of approach is always FS-DS and the difference between the top performing approach and the other approach (i.e., FS) is statistically significant across all data quality levels and for each level of data quality.

5. Conclusion

While many studies investigated feature selection and data sampling in bioinformatics separately, utilizing them together has received little attention. In this work, we compare three approaches for combining feature selection and data sampling (DS-FS-UnSam, DS-FS-Sam, and FS-DS). We also show the importance of alleviating class imbalance for classification problems on bioinformatics datasets. We employed three major forms of feature selection (feature ranking, filter-based subset selection, and wrapper-based feature selection) as well as a commonly used data sampling technique. We created three categories of datasets (High-Quality, Average-Quality, and Low-Quality) by injecting artificial class noise in a controlled fashion into ten gene-expression datasets which were first determined to be relatively free of noise. We build our final models using six different classification algorithms.

The experimental results demonstrate that paying attention to the order when utilizing both feature selection and data sampling and the dataset (whether unsampled or sampled) used for classification is extremely important in improving the performance of classification algorithms. We found that the best order to apply feature selection and data sampling is to employ feature selection followed by data sampling. This approach significantly improved the performance of all classifiers compared to the other approaches. All of these results are supported by ANOVA and Tukey's HSD tests. On the other hand, the results show that data sampling (in conjunction with feature selection) helped improve the classification performance even more compared to feature selection alone. Thus, we recommend alleviating class imbalance (e.g., by applying RUS) to achieve improved classification performance for bioinformatics classification problems. In particular, we recommend using FS-DS as the approach when learning from class imbalanced high dimensional bioinformatics datasets, regardless of any implication of noise or the classification algorithm that is going to be used. Furthermore, we recommend using FS-DS with feature rankers (especially ROC and WRS utilized with 100 features), as they showed superior classification performance compared to subset-based feature selection techniques.

Future research may involve conducting more experiments, using other classification algorithms as well as other learners within the wrapper, and considering other preprocessing techniques.

Acknowledgements

The authors would like to thank the anonymous reviewer for their constructive evaluation of this Book Chapter, and the various members of the Data Mining and Machine Learning Laboratory, Florida Atlantic University, for assistance with the reviews.

References

[1] Abu Shanab, A. and T.M. Khoshgoftaar. 2018. Is gene selection enough for imbalanced bioinformatics data. pp. 346–355. *In*: IEEE International Conference on Information Reuse and Integration (IRI), July 2018.

[2] Abu Shanab, A., T.M. Khoshgoftaar, R. Wald and J. Van Hulse. 2011. Comparison of approaches to alleviate problems with high-dimensional and class-imbalanced data. pp. 234–239. *In*: IEEE International Conference on Information Reuse and Integration (IRI), August 2011.

[3] Abu Shanab, A., T.M. Khoshgoftaar and R. Wald. 2012. Robustness of threshold-based feature rankers with data sampling on noisy and imbalanced data. pp. 92–97. *In*: Proceedings of the Twenty-Fifth International Florida Artificial Intelligence Research Society Conference.

[4] Al-Shahib, A., R. Breitling and D. Gilbert. 2005. Feature selection and the class imbalance problem in predicting protein function from sequence. Applied Bioinformatics 4(3): 195–203.

[5] Beer, D.G., S.L.R. Kardia, C.-C. Huang, T.J. Giordano, A.M. Levin, D.E. Misek, L. Lin, G. Chen, T.G. Gharib, D.G. Thomas, M.L. Lizyness, R. Kuick, S. Hayasaka, J.M.G. Taylor, M.D. Iannettoni, M.B. Orringer and S. Hanash. 2002. Gene-expression profiles predict survival of patients with lung adenocarcinoma. Nat. Med. 8(8): 816–824, Aug 2002 [Online]. Available: http://dx.doi.org/10.1038/nm733.

[6] Berenson, M.L., D.M. Levine and M. Goldstein. 1983. Intermediate Statistical Methods and Applications: A Computer Package Approach. Englewood Cliffs, New Jersey: Prentice-Hall.

[7] Berthold, M. and D.J. Hand (eds.). 2004. Intelligent Data Analysis, 2nd ed. Secaucus, NJ, USA: Springer-Verlag New York, Inc.

[8] Blagus, R. and L. Lusa. 2010. Class prediction for high-dimensional class-imbalanced data. BMC Bioinformatics 11(1): 523–539 [Online]. Available: http://www.biomedcentral.com/1471-2105/11/523.

[9] Blagus, R. and L. Lusa. 2012. Evaluation of smote for high-dimensional class-imbalanced microarray data. pp. 89–94. *In*: Machine Learning and Applications (ICMLA), 2012 11th International Conference on, Vol. 2, Dec 2012.

[10] Breiman, L. 2001. Random forests. Machine Learning 45(1): 5–32, Oct 2001 [Online]. Available: http://dx.doi.org/10.1023/A: 1010933404324.

[11] Breitling, R. and P. Herzyk. 2005. Rank-based methods as a non-parametric alternative of the t-statistic for the analysis of biological microarray data. Journal of Bioinformatics and Computational Biology 3(5): 1171–1189 [Online]. Available: http://www.worldscientific.com/doi/abs/10.1142/S0219720005001442.

[12] Caruana, R. and D. Freitag. 1994. Greedy attribute selection. pp. 28–36. *In*: Proceedings of the Eleventh International Conference on Machine Learning. Morgan Kaufmann.

[13] Chen, X.-w. and M. Wasikowski. 2008. Fast: a roc-based feature selection metric for small samples and imbalanced data classification problems. pp. 124–132. *In*: KDD '08: Proc. 14th ACM SIGKDD Int'l Conf. Knowledge Discovery and Data Mining. New York, NY, USA: ACM.

[14] Cristianini, N. and B. Schölkopf. 2002. Support vector machines and kernel methods: The new generation of learning machines. AI Mag. 23(3): 31–41, Sep. 2002 [Online]. Available: http://dl.acm.org/citation.cfm?id=765580.765585.

[15] Dittman, D.J., T.M. Khoshgoftaar, R. Wald and A. Napolitano. 2012. Determining the number of iterations appropriate for ensemble gene selection on microarray data. pp. 82–89. *In*: Machine Learning and Applications (ICMLA), 2012 11th International Conference on, Vol. 1, Dec. 2012.

[16] Dittman, D.J., T.M. Khoshgoftaar, R. Wald and A. Napolitano. 2013. Maximizing classification performance for patient response datasets. pp. 454–462. *In*: Tools with Artificial Intelligence (ICTAI), 2013 IEEE 25th International Conference on.

[17] Dittman, D.J., T.M. Khoshgoftaar, R. Wald and H. Wang. 2011. Stability analysis of feature ranking techniques on biological datasets. pp. 252–256. *In*: IEEE International Conference on Bioinformatics and Biomedicine (BIBM), November 2011.

[18] Domingos, P. and M. Pazzani. 1997. On the optimality of the simple bayesian classifier under zero-one loss. Mach. Learn. 29(2-3): 103–130, Nov. 1997 [Online]. Available: http://dx.doi.org/10.1023/A:1007413511361.

[19] Fraiman, R., A. Justel and M. Svarc. 2010. Pattern recognition via projection-based knn rules. Computational Statistics and Data Analysis 54(5): 1390–1403.

[20] Frenay, B. and M. Verleysen. 2014. Classification in the presence of label noise: A survey. Neural Networks and Learning Systems, IEEE Transactions on 25(5): 845–869, May 2014.

[21] Gordon, G.J., R.V. Jensen, L.-L. Hsiao, S.R. Gullans, J.E. Blumenstock, S. Ramaswamy, W.G. Richards, D.J. Sugarbaker and R. Bueno. 2002. Translation of microarray data into clinically relevant cancer diagnostic tests using gene expression ratios in lung cancer and mesothelioma. Cancer Research 62(17): 4963–4967 [Online]. Available: http://cancerres. aacrjournals.org/content/62/17/4963.abstract.

[22] Guo, X., Y. Yin, C. Dong, G. Yang and G. Zhou. 2008. On the class imbalance problem. pp. 192–201. *In*: Fourth International Conference on Natural Computation, 2008. ICNC '08, Vol. 4, October 2008.

[23] Guyon, I. and A. Elisseeff. 2003. An introduction to variable and feature selection. J. Mach. Learn. Res. 3: 1157–1182.

[24] Hall, M. 1997. Correlation-based Feature Selection for Machine Learning. Ph.D. dissertation, The University of Waikato, Hamilton, New Zealand.

[25] Hosmer, D.W. and S. Lemesbow. 1980. Goodness of fit tests for the multiple logistic regression model. Communications in Statistics-Theory and Methods 9(10): 1043–1069 [Online]. Available: http://www.tandfonline.com/doi/abs/10.1080/03610928008827941.

[26] Iizuka, N., M. Oka, H. Yamada-Okabe, M. Nishida, Y. Maeda, N. Mori, T. Takao, T. Tamesa, A. Tangoku, H. Tabuchi, K. Hamada, H. Nakayama, H. Ishitsuka, T. Miyamoto, A. Hirabayashi, S. Uchimura and Y. Hamamoto. 2003. Oligonucleotide microarray for prediction of early intrahepatic recurrence of hepatocellular carcinoma after curative resection. The Lancet 361(9361): 923–929.

[27] Inza, I., P. Larrañaga, R. Blanco and A.J. Cerrolaza. 2004. Filter versus wrapper gene selection approaches in DNA microarray domains. Artificial Intelligence in Medicine 31(2): 91–103 [Online]. Available: http://dx.doi.org/10.1016/j.artmed.2004.01.007.

[28] Jiang, Y., J. Lin, B. Cukic and T. Menzies. 2009. Variance analysis in software fault prediction models. pp. 99–108. *In*: Software Reliability Engineering, 2009. ISSRE '09. 20th International Symposium on, Nov. 2009.

[29] Khoshgoftaar, T.M., D. Dittman, R. Wald and A. Fazelpour. 2012. First order statistics based feature selection: A diverse and powerful family of feature seleciton techniques. pp. 151–157. *In*: 11th International Conference on Machine Learning and Applications (ICMLA), Vol. 2, Dec. 2012.

[30] Khoshgoftaar, T.M., A. Fazelpour, H. Wang and R. Wald. 2013. A survey of stability analysis of feature subset selection techniques. pp. 424–431. *In*: Information Reuse and Integration (IRI), IEEE 14th International Conference on, Aug 2013.

[31] Kohavi, R. 1995. A study of cross-validation and bootstrap for accuracy estimation and model selection. pp. 1137–1143. *In*: Proceedings of the 14th International Joint Conference on Artificial Intelligence—Volume 2, ser. IJCAI'95. San Francisco, CA, USA: Morgan Kaufmann Publishers Inc. [Online]. Available: http://dl.acm.org/citation. cfm?id=1643031.1643047.

[32] Kotsiantis, S., D. Kanellopoulos and P. Pintelas. 2006. Handling imbalanced datasets: A review. GESTS International Transactions on Computer Science and Engineering 30(1): 25–36.

[33] Liu, H., J. Li and L. Wong. 2002. A comparative study on feature selection and classification methods using gene expression profiles and proteomic patterns. Genome Informatics 13: 51–60.

[34] Mitchell, T.M. 1997. Machine Learning, 1st ed. New York, NY, USA: McGraw-Hill, Inc.

[35] Petricoin, E.F. III, A.M. Ardekani, B.A. Hitt, P.J. Levine, V.A. Fusaro, S.M. Steinberg, G.B. Mills, C. Simone, D.A. Fishman, E.C. Kohn and L.A. Liotta. 2002. Use of proteomic

patterns in serum to identify ovarian cancer. The Lancet 359(9306): 572–577 [Online]. Available: http://www.sciencedirect.com/science/article/pii/S0140673602077462.

[36] Provost, F. and T. Fawcett. 2001. Robust classification for imprecise environments. Mach. Learn. 42(3): 203–231 [Online]. Available: http://dx.doi.org/10.1023/A:1007601015854.

[37] Ramaswamy, S., K.N. Ross, E.S. Lander and T.R. Golub. 2003. A molecular signature of metastasis in primary solid tumors. Nature Genetics 33: 49–54.

[38] Seliya, N., T.M. Khoshgoftaar and J. Van Hulse. 2009. A study on the relationships of classifier performance metrics. pp. 59–66. *In*: 21st International Conference on Tools with Artificial Intelligence, November 2009.

[39] Shipp, M.A., K.N. Ross, P. Tamayo, A.P. Weng, J.L. Kutok, R.C. Aguiar, M. Gaasenbeek, M. Angelo, M. Reich, G.S. Pinkus, T.S. Ray, M.A. Koval, K.W. Last, A. Norton, T.A. Lister, J. Mesirov, D.S. Neuberg, E.S. Lander, J.C. Aster and T.R. Golub. 2002. Diffuse large B-cell lymphoma outcome prediction by gene-expression profiling and supervised machine learning. Nature Medicine 8(1): 68–74, Jan. 2002 [Online]. Available: http://dx.doi.org/10.1038/nm0102-68.

[40] Van Hulse, J., T.M. Khoshgoftaar and A. Napolitano. 2011. A comparative evaluation of feature ranking methods for high dimensional bioinformatics data. pp. 315–320. *In*: IEEE International Conference on Information Reuse and Integration (IRI), August 2011.

[41] Van Hulse, J. and T.M. Khoshgoftaar. 2009. Knowledge discovery from imbalanced and noisy data. Data & Knowledge Engineering 68(12): 1513–1542 [Online]. Available: http://www.sciencedirect.com/science/article/pii/S0169023X09001141.

[42] Van Hulse, J., T.M. Khoshgoftaar and A. Napolitano. 2007. Experimental perspectives on learning from imbalanced data. pp. 935–942. *In*: Proceedings of the 24th International Conference on Machine Learning, ser. ICML '07. New York, NY, USA: ACM [Online]. Available: http://doi.acm.org/10.1145/1273496.1273614.

[43] Van Hulse, J., T.M. Khoshgoftaar, A. Napolitano and R. Wald. 2012. Threshold-based feature selection techniques for high-dimensional bioinformatics data. International Journal of Network Modeling and Analysis in Health Informatics and Bioinformatics 1(1-2): 47–61 [Online]. Available: http://dx.doi.org/10.1007/s13721-012-0006-6.

[44] Visa, S. and A. Ralescu. 2005. Issues in mining imbalanced data sets—a review paper. pp. 67–73. *In*: Proc. 16th Midwest Artificial Intelligence and Cognitive Science Conf.

[45] Wald, R., T.M. Khoshgoftaar and A. Napolitano. 2013. Stability of filter- and wrapper-based feature subset selection. pp. 374–380. *In*: Tools with Artificial Intelligence (ICTAI), IEEE 25th International Conference on, Nov 2013.

[46] Wang, Y., I.V. Tetko, M.A. Hall, E. Frank, A. Facius, K.F.X. Mayer and H.W. Mewes. 2005. Gene selection from microarray data for cancer classification—a machine learning approach. Computational Biology and Chemistry 29(1): 37–46 [Online]. Available: http://www.sciencedirect.com/science/article/B73G2-4F92463-1/2/55dd1384ae9cb8b7c2909abc8afba4f8.

[47] Witten, I.H., E. Frank and M.A. Hall. 2011. Data Mining: Practical Machine Learning Tools and Techniques, 3rd ed. Burlington, MA: Morgan Kaufmann, January 2011.

[48] Zhu, X. and X. Wu. 2004. Class noise vs. attribute noise: A quantitative study. Artificial Intelligence Review 22(3): 177–210, Nov 2004 [Online]. Available: http://dx.doi.org/10.1007/s10462-004-0751-8.

Chapter 3

Big Data and Class Imbalance in Medicare Fraud Detection

Richard A Bauder* and Taghi M Khoshgoftaar

1. Introduction

The healthcare industry produces a vast array of information ranging from patient records to provider payment and claims data [53], [38]. This industry has and continues to embrace big data in order to become more efficient and productive [48]. Big data is characterized by its vastness with typically very granular datasets, that, when used with advanced analysis techniques, can lead to potentially meaningful conclusions. The use of big data is often seen as the best, and sometimes only, paradigm for future business success [47]. The incorporation of big data provides dense layers of interconnections and potentially meaningful information but is often modeled directly without much consideration for fundamental data processing and engineering. Because big data is available and machine learning techniques can readily handle these copious amounts of data, building models directly using the entire dataset, with minimal prior data analysis or preparation, appears to be increasingly common [44]. Even so, directly using all the available data may not always be the most prudent course of action. Another important real-world issue often found in big data is that of class imbalance, which occurs simply because of an uneven balance in the number of positive and negative cases, or binary class labels, in a dataset [45]. Areas such as medical insurance fraud, where

College of Engineering & Computer Science, Florida Atlantic University, Boca Raton, Florida.
Email: khoshgof@fau.edu
* Corresponding author: rbauder2014@fau.edu

there are considerably fewer instances of fraud versus normal activities, experience class imbalance. The amount of data in the healthcare field is rapidly increasing via sources such as electronic health records and insurance claims records [53].

Another aspect of this increasing amount of information is the rise of the elderly population in the U.S., due to advances in healthcare and an overall increase in standard of living [2]. The number of elderly individuals rose 28% from 2004 to 2015, versus an increase of just 6.5% for those under 65 years of age [3]. Thus, the upkeep and improvement in the health of this population becomes more important to the elderly and their family and friends. This increased healthcare need comes at a price and is usually managed by a healthcare insurance program. In particular, U.S. healthcare spending grew by 4.3% in 2016 totaling over $3.3 trillion [4], [32]. Clearly, these programs need to be affordable to the general populace, but program costs, along with the elderly population, continue to increase, which can financially cripple individuals and families [33]. Medicare is a U.S. government program that provides healthcare insurance and financial support for the elderly population, ages 65 and older, and other select groups of beneficiaries [5]. Note that this program contributes to 20% of the overall U.S. healthcare spending. Within the Medicare program, each covered medical procedure is codified for claims and payment purposes. The basic claims process entails a physician performing one or more procedures and then submitting a claim to Medicare for payment, rather than directly billing the patient, thus assigning the role of "middle man" to Medicare in this process. A claim is defined as a request for payment for benefits or services received by a beneficiary.

In order to keep healthcare affordable, programs need to keep medical-related costs low. One way to do this involves reducing fraud, waste, and abuse (FWA) [30]. Malicious or wasteful activities can lead to higher costs and the possibility of patients going without necessary medical care. Some examples of fraud and abuse involve billing Medicare for appointments the patient failed to keep, services rendered that were more complex than those actually performed, unnecessary medical services, submitting excessive charges for services, drugs, or supplies, and misusing claims codes (e.g., upcoding or unbundling). Aside, from these typical fraud and abuse descriptions in Medicare, improper payments can also indicate possible fraud or abuse. The term improper payments refers to payments made by the government to the wrong person, in the wrong amount, or for the wrong reason [1]. Thus, finding improper payments could be a way to detect possible fraud and abuse activities. Even so, it is important to note that not all improper payments are considered fraud and abuse, but rather are related to clerical or bookkeeping errors. The interested reader can find additional information on Medicare and healthcare-related fraud and abuse in [14], [24], [25], [60]. Unfortunately, fraud is all to prevalent within healthcare with about 10% of all U.S. medical claims being fraudulent [49], [55]. Medicare alone accounted for up to 20% ($705.9 billion) of the total U.S. healthcare spending in 2017 [4]. Therefore, from the FBI fraud estimate, the possible fraud losses (and potential loss recovery) could be up to $70 billion in the Medicare program alone. The group, Coalition Against Insurance Fraud [31], provides statistics on fraud and abuse found in the U.S. healthcare system. Some of the more salient statistics include the recovery of $29.4 billion to Medicare since 1997 by the Health Care Fraud and Abuse Control program, the exclusion of 1,662 individuals and entities from Medicare and Medicaid claims and payments, and a nearly five-fold increase in the recovery of proceeds (i.e., civil recoveries). Even with these successful recoveries, medical fraud continues to be very attractive to would-be perpetrators, adversely influencing healthcare costs and quality of service. Therefore, the detection of fraud with increased cost recovery is critical for the continued viability of the Medicare program.

Traditionally, to detect Medicare fraud, a limited number of auditors, or investigators, are responsible for manually inspecting thousands of claims, but only have enough time to look for very specific patterns indicating suspicious behaviors [52]. In this study, we provide two case studies to demonstrate the effects of class imbalance with big data on the detection of fraud in the Medicare dataset with LEIE fraud labels [37]. We use the following three different datasets, with provider payment and utilization information, released by the Centers for Medicare and Medicaid Services (CMS) [22]: (1) Medicare Provider Utilization and Payment Data: Physician and Other Supplier (Part B), (2) Medicare Provider Utilization and Payment Data: Part D Prescriber (Part D), and (3) Medicare Provider Utilization and Payment Data: Referring Durable Medical Equipment, Prosthetics, Orthotics, and Supplies (DMEPOS). We chose these parts of Medicare because they cover a wide range of possible provider claims, the information is presented in similar formats, and they are publicly available. Our study focuses on claims where the providers (e.g., physicians) determine what they will charge and bill for. The Part B, Part D, and DMEPOS datasets comprise key components of Medicare, which enables us to provide a comprehensive view of fraud in the Medicare program. Additionally, we create a combined dataset encompassing all provider claims across the three Medicare datasets. Information provided in these datasets includes the average amount paid for these services and other data points related to procedures performed, drugs administered, or supplies issued. The provided Medicare datasets do not have associated fraud labels for predicting possible fraud. We use the List of Excluded Individuals and Entities (LEIE) [46] dataset to generate fraud class labels (i.e., fraud or no fraud) for each provider to assess fraud detection capabilities of our baseline model and proposed improvement strategies. The LEIE contains all physicians who are excluded from practicing medicine for federally funded programs, such as Medicare.

The mapping of these LEIE fraud labels to each dataset indicates severe class imbalance. In the first case study, to address the issue of class imbalance, we create seven class distributions, or ratios, employing the random undersampling (RUS) technique and build Random Forest models for each distribution. For each of the models, we use 5-fold cross-validation repeated 10 times to reduce bias, assessing fraud detection performance using the Area Under the receiver operator characteristic Curve (AUC). The first case study indicates that the 90:10 (majority:minority) class distributions produces the best overall results. We clearly demonstrate that, in contrast to its prosaic use, the 50:50 class distribution does not produce the best results. Our research shows statistically significant class distribution differences and similarities in generating good fraud detection performance, as well as trends in class distribution model results. These results clearly show that the 50:50 (balanced) or 99:1 (imbalanced) class distributions have statistically similar fraud detection performance. Our second case study takes the 90:10 class distribution for fraud detection using the Part D, DMEPOS, and combined datasets. Overall, we show the value of RUS in improving fraud detection performance across the Medicare datasets, indicating good results with the Random Forest model. Our main contributions can be summarized as follows:

- Discuss a novel and robust Medicare data preparation approach.
- Detail our unique LEIE fraud labeling mapping methodology.
- Show that the commonly used 50:50 (balanced) class distribution does not produce top Medicare fraud detection results.
- Demonstrate class distribution results and trends that show significant differences in model performance for Medicare fraud detection.
- Show promising fraud detection results across several big data Medicare sources, leveraging the 90:10 RUS class distribution.

The rest of the paper is organized as follows. Section 2 discusses works related to the current research, focusing on class imbalance and Medicare-related fraud. We discuss the Medicare dataset and LEIE database, to include data preparation and fraud label mapping, in Section 3. In Section 4, we discuss the design of our experiment which includes class imbalance, the Random Forest learner, and performance metric. In Section 5, the results of our case studies are discussed. Finally, Section 6 summarizes our conclusions and future work.

2. Related Works

Our research compares and contrasts Medicare fraud detection performance using all the available data versus applying sampling to mitigate the effects of class imbalance. Therefore, we intentionally focus on any related works on Medicare fraud detection and/or class imbalance. Given this, there are relatively few studies on Medicare fraud detection, especially works utilizing the known provider exclusion database that take into account class imbalance.

A study by Ko et al. [43] uses only the 2012 Medicare data with a focus on the Urology specialty. The authors calculated the variability among Urologists, which indicated a possible savings of 9% due to provider utilization. Pande et al. [50] use 2012 Medicare data and exclusions from the LEIE database to assess who the Medicare fraud perpetrators are and what happens to them after they get caught. Interestingly, one of the authors' recommendations is to use predictive models to detect claims fraud. Khurjekar et al. [42] propose a two-step unsupervised approach to detecting fraud using the 2012 Medicare data. The authors first use the residuals from a multivariate regression model, with average payment as the dependent variable, to identify suspicious claims based on a residual threshold of $500. The second part of their approach incorporates these residuals using clustering to find fraudulent observations based on the average cluster distances. Another study by Sadiq et al. [54] employs the Patient Rule Induction Method (PRIM) based bump hunting method to identify anomalies in the 2014 Medicare data (Florida only). Their method is unsupervised and is used to narrow down the list of possibly fraudulent providers to be further investigated. In a preliminary study, Chandola et al. [19] use Medicare claims data and provider enrollment data from private sources to detect healthcare fraud. The authors employ several different techniques including social network analysis, text mining, and temporal analysis. Using features derived from the temporal analysis, the authors build a logistic regression model to detect known fraudulent cases using labeled data from the Texas Office of Inspector General's exclusion database, not the complete LEIE database. Moreover, details are limited with regards to data processing and mapping fraud labels to the Medicare data.

A two-step approach in detecting Medicare fraud, per provider type, is outlined in a paper by our research group [10]. The first step involves a multivariate regression model returning model residuals. These residuals are passed into a Bayesian probability model that produces the final probabilities indicating how likely it is that a particular value is fraudulent. We compared their method versus other common outlier detection methods, and found our method performed favorably. In [13], we provide an exploratory study predicting fraudulent providers using only the number of procedures performed by each physician, via a Multinomial Naive Bayes model. If the predicted provider type does not match what is expected, then this provider is performing outside of normal practice patterns and should be investigated. In [8], we use multivariate regression to establish a baseline for expected Medicare payments, per provider type. This baseline is then used as the normative case in which to compare the actual payment amounts, with deviations flagged as outliers. A two-step approach in detecting Medicare fraud, per provider type, is outlined in [10]. Another previous research study [9] involves a preliminary study that compares several supervised and unsupervised methods to detect 2015 Medicare Part B fraud. In this study, we detect fraud with supervised (Gradient Boosted Machine, Random Forest, Deep Neural Network, and Naive Bayes), unsupervised (autoencoder, Mahalanobis distance, KNN, and LOF), and hybrid (multivariate regression and Bayesian probability) machine learning approaches. Supervised methods performed better than unsupervised or hybrid approaches, with results fluctuating based on the sampling technique used and Medicare provider type.

Branting et al. [16] create a graph of providers, prescriptions, and procedures using the 2012 to 2014 Medicare data and LEIE exclusion labels. The authors use two algorithms where one calculates the similarity to known fraud and non-fraud providers, and the other estimates fraud risk via shared practice locations. To address class imbalance, the authors kept 12,000 excluded providers and randomly selected 12,000 non-excluded providers, using only a 50:50 class distribution. A decision tree model was built using 11 graph-based features and 10-fold cross-validation with no repeats.

To the best of our knowledge, our work is one of the only Medicare fraud detection studies, to provide such a robust experiment to assess the impacts of using big data, with severe class imbalance. To support our assertions, we use the Random Forest model to demonstrate the significant improvements by employing sampling and suggest the best class distributions while debunking the common usage of the 50:50 distribution. Moreover, contrary to the related works, we provide a comprehensive and fair experimental design using 5-fold cross-validation with 10 repeats for each class distribution, as well as statistical significance testing.

3. Data

To effectively demonstrate Medicare fraud detection performance, we use three publicly available Medicare provider claims data from the Centers for Medicare and Medicaid Services (CMS) [22]. Additionally, we create a combined dataset incorporating each of these three big Medicare datasets. In these datasets, each provider or physician is denoted by his or her unique National Provider Identifier (NPI) [23] for each medical claim item. The Medicare dataset contains a number of features, such as the average amount submitted, billed, and paid by Medicare, and the number of procedures performed. Note that the Medicare claims information is recorded after claims payments were made [26] and with that, we do not make any modifications and assume that this dataset was appropriately recorded and cleansed by CMS. By using these claims datasets, we demonstrate the detection of fraudulent behaviors at the provider-level. This implies a single provider with a single procedure per Medicare claim. Before implementing the CMS data in research, it is important to understand each dataset and how to manipulate and leverage it in the most efficient and effective way [11].

The Medicare Provider Utilization and Payment Data: Physician and Other Supplier (Part B) dataset, from 2012 to 2015, outlines information about physicians and the procedures they perform [28]. Each physician is denoted by his or her NPI and each procedure is labeled by its Healthcare Common Procedure Coding System (HCPCS) code [21]. The Part B data is aggregated (grouped by) the following: (1) NPI of the performing provider, (2) HCPCS code for the procedure or service performed, and (3) the place of service which is either a facility (F) or non-facility (O), such as a hospital or office, respectively. Some physicians can perform the same procedure (i.e., have the same HCPCS code) at both a facility and an office. Additionally, there are a few cases for which a physician is labeled as multiple physician types (or specialties), such as Internal Medicine and Cardiology. The Part B data, per year, is organized where each row contains the physician's NPI and provider type (along with all non-changing physician information, such as name and gender) corresponding to one HCPCS code and further split by place of service (Office or Facility). Given this organization, all the procedure information corresponds to these four attributes. Therefore, for each physician, there are as many rows as unique combinations of NPI, Provider Type, HCPCS code, and place of service. For example, if a physician (NPI = 1003000126) has claimed 20 different procedures and three of them were conducted at both an office and facility (while the other 17 were conducted at one place), there would be 23 rows for this physician (assuming this physician is labeled as only one provider type).

The Medicare Provider Utilization and Payment Data: Part D Prescriber (Part D) dataset, from 2013 to 2015, outlines information about physicians, as well as information pertaining to the prescription drugs they administer under the Medicare Part D Prescription Drug Program [27]. Each physician is denoted by his or her NPI and each drug is labeled by its brand and generic name. The Part D data is aggregated (grouped by) the following: (1) the NPI of the prescriber, (2) the drug name (brand name in the case of trademarked drugs) and generic name (according to CMS documentation). As with the Part B data, there are a few cases where a physician can be labeled as multiple physician types, such as: internal medicine and cardiology. The Part D data, per year, is organized where each row contains the physician's NPI and provider type (along with all non-changing physician information) corresponding to one drug name along with all the drug information corresponding to these three attributes. Therefore, for each physician, there are as many rows as unique combinations of NPI, Provider type and drug name. For example, if a physician (NPI = 1003000126) has prescribed 20 different drugs, there would be 20 rows for this physician (assuming this physician is labeled as one physician type).

The Medicare Provider Utilization and Payment Data: Referring Durable Medical Equipment, Prosthetics, Orthotics and Supplies (DMEPOS) dataset, from 2013 to 2015, outlines information about physicians, as well as information pertaining to the DMEPOS products and services provided [29]. Each physician is denoted by his or her NPI and each product/service is labeled by its HCPCS code. The DMEPOS data is aggregated (grouped by) the following: (1) NPI of the performing provider, (2) HCPCS code for the procedure or service performed by the DMEPOS supplier, and (3) the supplier rental indicator (value of either 'Y' or 'N') derived from DMEPOS supplier claims (according to CMS documentation). Some physicians place orders for the same DMEPOS equipment (i.e., with the same HCPCS code), as both rental and non-rental. Additionally, there are also a few cases where a physician can be labeled as multiple physician types. The DMEPOS data, per year, is organized where each row contains the physician's NPI and provider type (along with all non-changing physician information) corresponding to one HCPCS code and further split by rental status (yes or no) and all the procedure information corresponding to these four attributes. Therefore, for each physician, there are as many rows as unique combinations of NPI, Provider type, HCPCS code, and rental_indicator. As an example, if a physician (NPI = 1003000126) has claimed 20 different procedures and three of them were issued as both a rental and non-rental (while the other 17 were issued as one), there would be 23 rows for this physician (assuming this physician is labeled as one physician type). For additional clarity and insight into the Medicare data, Tables 1, 2, and 3 depict sample excerpts, from the Internal Medicine provider type or specialty, from each of the three Medicare datasets used in this paper (with obfuscated NPI values of '1111111111').

In combining each of the individual years for the 2012 (or 2013) to 2015 Medicare datasets, we matched features and excluded those that did not match across all years. For instance, with the 2012 Part B dataset, the standard deviations for charges and payments are available but discontinued for the later years and were not included in the final dataset. Additionally, we create a combined dataset incorporating information from all three Medicare datasets. Our assumption is that there is no reliable way to know within which part of Medicare a physician/provider has or will commit fraud. Therefore, joining

npi	...	nppes_provider_gender	...	provider_type	...	place_of_service	hcpcs_code	...	line_srvc_cnt	bene_unique_count	...	average_submitted_chrg_amt	...
1111111111	...	M	...	Internal Medicine	...	F	99217	...	23	23	...	328.00000	...
1111111111	...	M	...	Internal Medicine	...	F	99219	...	18	18	...	614.00000	...
1111111111	...	M	...	Internal Medicine	...	F	99221	...	59	58	...	333.28814	...
1111111111	...	M	...	Internal Medicine	...	F	99231	...	38	18	...	100.84211	...
1111111111	...	M	...	Internal Medicine	...	F	99232	...	1117	481	...	200.93196	...
1111111111	...	M	...	Internal Medicine	...	F	99291	...	21	13	...	633.80952	...

Table 1: Part B dataset sample.

npi	...	specialty_description	...	drug_name	...	total_claim_count	...	total_day_supply	total_drug_cost	...	total_claim_count_ge65	ge65_suppress_flag	...
1111111111	...	Internal Medicine	...	AMLODIPINE BESYLATE	...	27	...	990	120.01	...	NA	#	...
1111111111	...	Internal Medicine	...	ATORVASTATIN CALCIUM	...	15	...	450	188.85	...	NA	*	...
1111111111	...	Internal Medicine	...	AZITHROMYCIN	...	16	...	87	139.24	...	NA	#	...
1111111111	...	Internal Medicine	...	CEPHALEXIN	...	12	...	96	76.09	...	NA	#	...
1111111111	...	Internal Medicine	...	CIPROFLOXACIN HCL	...	15	...	114	119.36	...	NA	#	...
1111111111	...	Internal Medicine	...	CLOPIDOGREL	...	24	...	780	205.46	...	NA	#	...
1111111111	...	Internal Medicine	...	FUROSEMIDE	...	12	...	360	34.83	...	NA	*	...
1111111111	...	Internal Medicine	...	HYDRALAZINE HCL	...	14	...	375	249.54	...	14		...

Table 2: Part D dataset sample.

the Part B, Part D, and DMEPOS datasets can potentially better represent a provider's claims, from procedures and drugs to equipment. This is because the combined dataset has a larger number of features from which machine learning algorithms can detect fraud.

None of the aforementioned CMS-provided Medicare datasets include fraud labels, or other indicators for possible fraudulent claims. In order to obtain labels indicating fraudulent providers, we incorporate excluded providers from the LEIE database [46]. The LEIE database is updated monthly, so for our study, we used the LEIE dataset released on January 3, 2018. The provider exclusions are categorized by various rule numbers, which indicate the severity and minimum exclusion period. As seen in Table 4, we selected only mandatory exclusions (not permissive exclusions), indicating more severe convictions and/or revocations. Note that, in generating the labels for model building, we assume that excluded providers are considered fraudulent and those not on the exclusion list are non-fraudulent. Unfortunately, the LEIE does not contain an NPI number for most of the available providers. Even so, in order to maintain the most accurate fraud label mappings, we only use provider NPIs and exclude any providers without a NPI number. Additionally, we only included features found in all four years. For instance, in 2012 the standard deviations for charges and payments are available but discontinued for the later years. More specifically, in combining the 2012 to 2015 Medicare datasets with exclusion labels, we cross-referenced NPI numbers in the Medicare data and LEIE database, to match any providers with past or current exclusions.

As mentioned, the Medicare data contains annual claims information by provider and specific procedure performed, as well as the place this service was performed, whereas the LEIE database only contains information for the provider and not any particular procedure or location. Currently, there is no known publicly available data source with fraud labels by provider and by each procedure performed. In order to account for this discrepancy and correctly map LEIE exclusion labels to the Medicare dataset, we decided to aggregate the Medicare data at the provider- or NPI-level. After filtering the Medicare data based on the drug indicator, removing any prescription information, and Medicare participation, we grouped the data by the specialty (also known as the provider type), NPI, and gender and aggregated across all procedures and places of services. In order to avoid too much information loss due to the aggregation, we generated additional numeric features from the original five to include the mean, sum, median, standard deviation, minimum, and maximum. Additionally, we retained the specialty and gender categorical features. In order to build our model with a mixture of numerical and categorical features, we employed one-hot encoding. This method uses the categorical values to generate dummy features with binary values which indicate the presence of

Table 3: DMEPOS dataset sample.

REFERRING_NPI	...	REFERRING_PROVIDER_TYPE	...	HCPCS_CODE	...	SUPPLIER_RENTAL_INDICATOR	NUMBER_OF_SUPPLIERS	...	NUMBER_OF_SUPPLIER_CLAIMS	...	AVG_SUPPLIER_SUBMITTED_CHARGE	...
1111111111	...	Internal Medicine	...	E0431	...	Y	6	...	51	...	48.8546154	...
1111111111	...	Internal Medicine	...	E1390	...	Y	6	...	85	...	251.0091861	...

Table 4: LEIE exclusion rules.

Rule Number	Description
1128(a)(1)	Conviction of program-related crimes.
1128(a)(2)	Conviction for patient abuse or neglect.
1128(a)(3)	Felony conviction due to healthcare fraud.
1128(b)(4)	License revocation or suspension.
1128(c)(3)(g)(i)	Conviction of 2 mandatory offenses.
1128(c)(3)(g)(ii)	Conviction on 3+ mandatory offenses.

this variable, assigning a value of one if present otherwise zero, versus all other dummy features. This translates each of the original categorical values into distinct binary features. Table 5 describes each of the Medicare features from which the aggregated dataset is generated, as well as the categorical and class (exclusion) features, for each dataset.

After the Medicare data NPI-level aggregation, we map the LEIE-excluded providers as fraud labels. From the described mandatory exclusions, only rules with 5-year minimum exclusion periods were found in the LEIE data. Thus, each exclusion period has a 5-year length of time. We take the provided exclusion date (i.e., the start of the exclusion period) and add 5 years to get the end date of the exclusion period. We then compare the start and end exclusion dates to any listed waiver or reinstatement dates. The assumption is that if there is a waiver or reinstatement date, then any activities on or after this date are no longer considered fraudulent. We updated the end date of the exclusion period based on the wavier and reinstatement comparisons. For example, if the exclusion end date is 2016/03/12 and the waiver date is 2014/02/01, then the updated exclusion end date is 2014/02/01. Each provider in the LEIE database has start and updated end exclusion dates that can be used during the integration with the Medicare data. We merge the Medicare and LEIE datasets using NPI as the key and create an *exclusion* feature to store fraud labels. Labels are assigned as fraud if a provider's Medicare year is less than the exclusion end date (for which we use the year because the Medicare dataset only contains years), otherwise *exclusion* is kept as non-fraud. In order to avoid too few or too many fraud labels, we round the new exclusion end date to the nearest year based on the month. So, if the month is greater than 6, then the exclusion end year is increased to the following year, otherwise the current year is used. In this way, partial years are addressed with the assumption that if an exclusion end date occurs during the latter part of a year, the majority of that year can be assumed as fraud. Otherwise, if very little of the year is before the exclusion end date, then we assume the provider claims in that year are not fraudulent. This labeling includes both the exclusion period and the period prior to the start of the exclusion. The rationale for keeping the former is that claims made during the exclusion period are improper payments and could be considered fraudulent per the federal False Claims Act (FCA) [30]. The latter is kept as it indicates fraudulent behaviors leading up to that provider being put on the LEIE. This process to map the LEIE exclusion labels takes into account overlapping exclusion and Medicare claims periods to avoid mapping unnecessary fraud labels.

Table 6 summarizes the Medicare, NPI-level aggregated datasets with fraud labels. The number of fraud labels across datasets clearly shows the severe class imbalance. After the data aggregation and fraud label mapping, only the NPI feature is not used to build or test the models, but rather for

Table 5: Description of medicare dataset features.

Dataset	Feature	Description	Type
Part B	npi	Unique provider identification number	Categorical
	provider_type	Medical provider's specialty (or practice)	Categorical
	nppes_provider_ gender	Provider's gender	Categorical
	line_srvc_cnt	Number of procedures/services the provider performed	Numerical
	bene_unique_cnt	Number of distinct Medicare beneficiaries receiving a service	Numerical
	bene_day_srvc_cnt	Number of distinct Medicare beneficiary/per day services	Numerical
	average_submitted_ chrg_amt	Average of the charges that the provider submitted for a service	Numerical
	average_medicare_ payment_amt	Average payment made to a provider per claim for a service	Numerical
Part D	npi	Unique provider identification number	Categorical
	specialty_description	Medical provider's specialty (or practice)	Categorical
	bene_count	Number of distinct Medicare beneficiaries receiving the drug	Numerical
	total_claim_count	Number of drug the provider administered	Numerical
	total_30_day_fill_ count	Number of standardized 30-day fills	Numerical
	total_day_supply	Number of day's supply	Numerical
	total_drug_cost	Cost paid for all associated claims	Numerical
DMEPOS	referring_npi	Unique provider identification number	Categorical
	referring_provider_ type	Medical provider's specialty (or practice)	Categorical
	referring_provider_ gender	Provider's gender	Categorical
	number_of_suppliers	Number of suppliers used by provider	Numerical
	number_of_supplier_ beneficiaries	Number of beneficiaries associated by the supplier	Numerical
	number_of_supplier_ claims	Number of claims submitted by a supplier from a referring order	Numerical
	number_of_supplier_ services	Number of services/products rendered by a supplier	Numerical
	avg_supplier_ submitted_charge	Average payment submitted by a supplier	Numerical
	avg_supplier_ medicare_pmt_amt	Average payment awarded to suppliers	Numerical
All	exclusion	Fraud labels mapped from the LEIE dataset	Categorical

Table 6: Summary of medicare datasets.

Dataset	Non-Fraud Instances (#)	Fraud Instances (#)	Fraud Instances (%)
Part B	3,691,146	1,409	0.038
Part D	2,098,715	1,018	0.048
DMEPOS	862,792	635	0.074
Combined	759,267	473	0.062

identification purposes. The use of any remaining variables or derived features, along with applying feature engineering approaches, is left as future work.

4. Experimental Design

In this section, we detail our experiment methodology. We discuss class imbalance, the Random Forest model, cross-validation, performance metrics, and significance testing.

4.1 Class Imbalance

In our study, we employ RUS to mitigate issues arising from the class imbalance problem [18], [56]. Due to the severe class imbalance between fraud and non-fraud labels, a model will tend to focus on the majority class (i.e., the class with the majority of instances) and misrepresent the minority class. In our case, the non-fraud labels are the majority class and the fraud labels are the minority class, as well as the class of interest in our study. The use of data sampling changes the class distribution of the training instances by increasing the representation of the minority class, thus helping to improve model performance. There are two basic sampling methods: oversampling and undersampling. Oversampling is a method for altering the distribution of classes in a dataset by adding instances to the minority class, whereas undersampling removes samples from the majority class. Of course, as with most methods, there are disadvantages. With undersampling, the main disadvantage is discarding potentially useful information. Oversampling, because it duplicates existing minority class instances, can increase the likelihood of overfitting [20]. Oversampling can also increase processing time by increasing the overall size of the data. Our choice to use only RUS is further supported in [7], [12], [36], [41], [59].

For our experiments, we generate the following class distributions (majority:minority): 99.9:0.1, 99:1, 95:5, 90:10, 75:25, 65:35, and 50:50. Most of these class ratios retain a reasonable amount of the majority class and reduce loss of information relative to the minority class. In order to mitigate some of the potential majority class information loss using RUS, we repeat the

Table 7: RUS class distribution sample size.

Fraud		Non-fraud		
%	#	%	#	Total
0.1	1,409	99.9	1,407,591	1,409,000
1	1,409	99	139,491	140,900
5	1,409	95	26,771	28,180
10	1,409	90	12,681	14,090
25	1,409	75	4,227	5,636
35	1,409	65	2,617	4,026
50	1,409	50	1,409	2,818

sampling process 10 times for each class distribution. This effectively helps to reduce bias due to poor random draws and better represent the majority class through the use of different random samples. Table 7 summarizes the number of instances for the full dataset and each of the RUS datasets. This indicates that as the minority class (fraud) percentage increases, the representation of the majority class (non-fraud) decreases relative to the number of instances in the full dataset. Note that the number of fraud instances is always the same, regardless of percentage.

4.2 *Random Forest*

In order to assess the performance using all the data versus the RUS datasets, we employ a Random Forest (RF) model. We selected the RF model because of its good classification performance, which has been shown to be superior to many other classifiers on a wide variety of datasets with or without class imbalance [34], [40]. Random Forest is an ensemble method in which multiple unpruned decision trees are built and a final classification is made by combining the results from the individual trees [17]. The algorithm creates random datasets using sampling with replacement to train each of the decision trees. At each node within a tree, RF chooses the most discriminating feature between the classes using entropy and information gain. Entropy can be seen as the measure of impurity or uncertainty of attributes, and information gain is a means to find the most informative attribute. Thus, the goal is to minimize entropy and maximize information gain with attribute selection. Additionally, RF performs random feature subspace selection, at each node of a tree, where a subset of m features are considered for the decision at that node. As seen in Figure 1, to classify a new instance X, pass X down each one of the N trees in the forest. Each tree gives a classification for this new instance. The forest then chooses the classification which has the majority out of N votes.

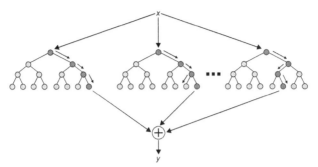

Fig. 1: Random forest classification process.

In our experiment, we use Weka [61] to build each RF model with 100 trees (annotated as RF100).

4.3 Cross Validation

We use k-fold cross-validation to evaluate the performance of the models. With this method, the model is trained and tested k times, where each time it is trained on $k - 1$ folds and tested on the remaining fold. This is to ensure that all data are used in the classification. More specifically, we use stratified cross-validation [61] which tries to ensure that each class (i.e., fraud or non-fraud) is approximately equally represented across each fold. In our study, we employ 5-fold cross-validation. Moreover, to further reduce bias due to bad random draws and to better represent the claims data, we repeat the 5-fold cross-validation process 10 times and average the scores to get the final results. Incorporating repeats allows for different randomly selected instances of the majority class to be used for each cross-validation step, thus providing a more representative sample of the non-fraud instances.

4.4 Performance Metric

Our RF100 model is a two-class classifier predicting fraud or no fraud instances. The model's accuracy can be represented by a confusion matrix consisting of information about actual and predicted classifications returned by a model. We use the AUC performance metric which is composed of values derived from the confusion matrix to assess a model's fraud detection performance [15], [57]. AUC is a popular measure of model performance, providing a general idea of predictive potential of a binary classifier. The receiver operating characteristic curve is used to characterize the trade-off between true positive (TP) rate, also known as recall or sensitivity, $(\frac{TP}{TP + FN})$ and false positive (FP) rate $(\frac{FP}{FP + TN})$, where FN is false negative and TN is true negative. This curve depicts a learner's performance across all classifier decision thresholds. AUC is a single value that ranges from 0 to 1, where a perfect classifier results in an AUC of 1. Additionally, due to the class imbalance in the Medicare data, we consider AUC a good means to assess fraud detection performance [39]. Note that we do not use other confusion matrix metrics directly, such as sensitivity and specificity, because a single discriminate classifier threshold of 0.5 is used to discriminate positive and negative classes. This single naive threshold is not appropriate for assessing model performance using highly imbalanced data.

4.5 Significance Testing

Hypothesis testing is performed to demonstrate the statistical significance of the results using ANalysis Of VAriance (ANOVA) [35] and *post hoc* analysis via Tukey's Honestly Significant Different (HSD) test [58]. ANOVA is a statistical test determining whether the means of several groups (factors) are equal. The Tukey's HSD test finds means of a factor that are significantly different from each other, comparing all possible pairs of means similar to a *t*-test. Differences are grouped by assigning letters, with pairs that do not share a common letter indicating significantly different results.

5. Results and Discussion

In our first case study, we evaluate fraud detection improvement through the systematic application of RUS, which reduces the adverse effects caused by class imbalance. We show that a 50:50 class distribution, which is typically used for many applications and generally has low model performance losses [51], is not the best ratio for Medicare fraud detection. The lower 50:50 distribution performance can be attributed to the small number of majority class instances, which may make it more difficult for a model to discriminate between fraud and non-fraud instances. This results in the misclassification of non-fraud instances as fraud instances, which increases the false positive rate and decreases overall model performance [6]. Figure 2 shows the trend of AUC scores across each of the minority class distributions. This trend depicts a decrease in scores, particularly with below a 1% minority class distribution.

Table 8 shows all of the AUC values for each class distribution, with the 90:10 class distribution producing the best overall average results with a low standard deviation. Even so, the difference in AUC scores between the top six class distributions is relatively small, so additional statistical significance testing is performed. Table 9a shows the results of a one-factor ANOVA with class distribution being significant at a 0.05 significance level. In order to get further details on the differences within the class distribution factor, we perform

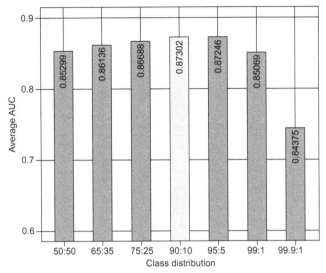

Fig. 2: RF100 AUC results by class distribution.

Table 8: RF100 AUC performance results by class distribution.

Class Distribution	Mean	Median	Standard Deviation	Minimum	Maximum
50:50	0.85299	0.85308	0.00350	0.84825	0.86045
65:35	0.86136	0.86109	0.00307	0.85786	0.86816
75:25	0.86688	0.86696	0.00255	0.86264	0.87122
90:10	0.87302	0.87266	0.00284	0.86876	0.87705
95:5	0.87246	0.87351	0.00403	0.86477	0.87809
99:1	0.85069	0.84882	0.00575	0.84422	0.86023
99.9:0.1	0.74375	0.74253	0.00707	0.73184	0.75907

a Tukey's HSD test. The difference in class distributions is shown in Table 9b. The 90:10 distribution has the highest AUC value which is significantly better than the remaining distributions. The 95:5 class distribution is the second best, in terms of fraud detection performance, and in a lower group than the 90:10 distribution. The 75:25, and 65:35 class distributions have some group overlap indicating little difference in fraud detection performance.

Interestingly, the 50:50 (balanced) and 99:1 (imbalanced) class distributions are in the same group, thus differences in performance between the two are statistically insignificant. Therefore, selecting the commonly used 50:50 distribution is not better than using 99:1 ratio which is highly imbalanced. Based on our results, we recommend using the 90:10 class distribution which has the best performance and is significantly better than the

Table 9: One-factor ANOVA for class distribution, with Tukey's HSD results.

	Df	Sum Sq	Mean Sq	F value	Pr (> F)
Distribution	6	0.12628	0.021047	1088	< 2e-16
Residuals	63	0.00122	0.000019		

(a) One-factor ANOVA results.

Group	Class Distribution	AUC
a	90:10	0.87302
ab	95:5	0.87246
bc	75:25	0.86688
c	65:35	0.86136
d	50:50	0.85299
d	99:1	0.85069
e	99.9:0.1	0.74375

(b) Tukey's HSD class distribution results.

other class distributions. Furthermore, this distribution retains a reasonable number of majority class instances providing a good representation of the majority (non-fraud) class, unlike the 50:50 class distribution. In addition to demonstrating that the 90:10 distribution has the best performance, our results suggest that there is a class distribution threshold where performance begins to decrease sharply. From Table 9b and Figure 2, we notice a possible threshold below the 99:1 class distribution, where using the 99.9:01 distribution is significantly worse.

Given the top performance of the 90:10 class distribution with Medicare Part B, in the second case study we demonstrate that the use of RUS produces good fraud detection results with other big Medicare datasets. In Figure 3, we show several violin plots with associated point scatter. The violin shape indicates the distribution shape of the data, where wider sections represent a higher probability that members of the population will take on the given value and the skinnier sections represent a lower probability. From these plots, we note that each of the datasets produces good results when using RF100 and the 90:10 class ratio. In particular, the combined dataset, with its larger feature space, exhibits the highest average model performance. This indicates that the added information, as well as the interactions between Medicare provider claims, increases the detection of possible fraudulent activities.

As with the Part B only results, we provide significance testing as seen in Table 10. The one-factor ANOVA, with the factor Dataset, indicates significant results at a 95% confidence interval. The Tukey's HSD results confirm the superiority of the results using the combined dataset, followed

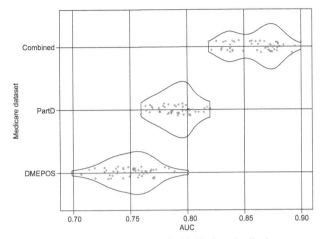

Fig. 3: RF100 AUC results for 90:10 class distribution.

Table 10: One-factor ANOVA for dataset, with Tukey's HSD results.

	Df	Sum Sq	Mean Sq	F value	Pr (> F)
Dataset	2	0.30856	0.15428	417.8	< 2e-16
Residuals	147	0.05428	0.00037		

(a) One-factor ANOVA results.

Group	Dataset	AUC
a	Combined	0.85989
b	PartD	0.79088
c	DMEPOS	0.74998

(b) Tukey's HSD class distribution results.

by Part D and DMEPOS. From the results in both case studies, we show that RUS is effective in increasing model performance for Medicare fraud detection, with the use of a slightly imbalanced class distribution exhibiting the best performance across all big Medicare datasets.

6. Conclusion

The use of big data from sources such as Medicare is being leveraged to improve patient care and to help detect fraud. Medicare fraud continues to be problematic for its beneficiaries and the U.S. economy, negatively impacting the ability of the Medicare program to provide effective and affordable care. Thus, it is critical to have effective fraud detection methods. In response to this concern, CMS has made available several large Medicare claims datasets for public use. Overall, in our study, we demonstrated the effectiveness of RUS in increasing fraud detection performance. We detail a unique method for processing the Medicare data and integrating the LEIE fraud labels. We then compare Medicare fraud detection performance in two case studies. In the first case study, with the Part B data only, we use seven different RUS class distributions. We build and test RF100 models using 5-fold cross-validation with 10 repeats assessing model performance using AUC. In the second case study, we use the best RUS class distribution and apply this to two other Medicare big datasets, as well as a combined dataset.

Our results indicate that the best class distribution is 90:10 with the worst results coming from the 99.9:0.01 distribution. We also showed that the performance of the commonly used ratio of 50:50 (balanced) is indistinguishable from the 99:1 class distribution. This indicates that the 50:50 distribution should not be used as the *de facto* standard for class imbalance in Medicare fraud detection. Moreover, we show, using the 90:10 class distribution, that we can build effective fraud detection models across other Medicare big datasets. Overall, we recommend using the 90:10 class distribution which indicates the best fraud detection performance. Furthermore, we noticed a possible threshold at the 99:1 class distribution, where any positive class representation below this performs significantly worse. Future work will include additional learners, as well as other performance metrics such as F-score and G-measure.

Acknowledgment

We would like to thank the reviewers in the Data Mining and Machine Learning Laboratory at Florida Atlantic University. Additionally, we acknowledge partial support by the NSF (CNS-1427536). Opinions, findings, conclusions, or recommendations in this paper are the authors' and do not reflect the views of the NSF.

References

[1] Improper Payments Elimination and Recovery Act of 2010. [Online]. Available: https://obamawhitehouse.archives.gov/sites/default/files/omb/financial/_improper/PL_111-204.pdf.

[2] How Growth of Elderly Population in US Compares With Other Countries. 2013. [Online]. Available: http://www.pbs.org/newshour/rundown/how-growth-of-elderly-population-in-us-compares-with-other-countries/.

[3] Profile of Older Americans: 2015. 2015. [Online]. Available: http://www.aoa.acl.gov/Aging_Statistics/Profile/2015/.

[4] National Health Expenditures 2017 Highlights. 2017. [Online]. Available: https://www.cms.gov/research-statistics-data-and-systems/statistics-trends-and-reports/nationalhealthexpenddata/downloads/highlights.pdf.

[5] US Medicare Program. 2017. [Online]. Available: https://www.medicare.gov.

[6] Batista, G.E., R.C. Prati and M.C. Monard. 2004. A study of the behavior of several methods for balancing machine learning training data. ACM Sigkdd Explorations Newsletter 6(1): 20–29.

[7] Bauder, R.A., T.M. Khoshgoftaar and T. Hasanin. 2018. An empirical study on class rarity in big data. pp. 785–790. *In*: 2018 17th IEEE International Conference on Machine Learning and Applications (ICMLA), Dec 2018.

[8] Bauder, R.A. and T.M. Khoshgoftaar. 2016. A novel method for fraudulent medicare claims detection from expected payment deviations (application paper). pp. 11–19. *In*: Information Reuse and Integration (IRI), 2016 IEEE 17th International Conference on, IEEE.

[9] Bauder, R.A. and T.M. Khoshgoftaar. 2017. Medicare fraud detection using machine learning methods. pp. 858–865. *In*: Machine Learning and Applications (ICMLA), 2017 16th IEEE International Conference on, IEEE.

[10] Bauder, R.A. and T.M. Khoshgoftaar. 2017. Multivariate outlier detection in medicare claims payments applying probabilistic programming methods. Health Services and Outcomes Research Methodology 17(3-4): 256–289.

[11] Bauder, R.A. and T.M. Khoshgoftaar. 2018. A survey of medicare data processing and integration for fraud detection. pp. 9–14. *In*: Information Reuse and Integration (IRI), 2018 IEEE 19th International Conference on, IEEE.

[12] Bauder, R.A., T.M. Khoshgoftaar and T. Hasanin. 2018. Data sampling approaches with severely imbalanced big data for medicare fraud detection. *In*: Tools with Artificial Intelligence (ICTAI), 2018 IEEE 30th International Conference on, IEEE.

[13] Bauder, R.A., T.M. Khoshgoftaar, A. Richter and M. Herland. 2016. Predicting medical provider specialties to detect anomalous insurance claims. pp. 784–790. *In*: Tools with Artificial Intelligence (ICTAI), 2016 IEEE 28th International Conference on, IEEE.

[14] Bauder, R.A., T.M. Khoshgoftaar and N. Seliya. 2017. A survey on the state of healthcare upcoding fraud analysis and detection. Health Services and Outcomes Research Methodology 17(1): 31–55.

[15] Bekkar, M., H.K. Djemaa and T.A. Alitouche. 2013. Evaluation measures for models assessment over imbalanced data sets. Journal of Information Engineering and Applications 3(10).

[16] Branting, L.K., F. Reeder, J. Gold and T. Champney. 2016. Graph analytics for healthcare fraud risk estimation. pp. 845–851. *In*: Advances in Social Networks Analysis and Mining (ASONAM), 2016 IEEE/ACM International Conference on, IEEE.

[17] Breiman, L. 2001. Random forests. Machine Learning 45(1): 5–32, Oct 2001 [Online]. Available: http://dx.doi.org/10.1023/A:1010933404324.

[18] Brennan, P. 2012. A comprehensive survey of methods for overcoming the class imbalance problem in fraud detection. Institute of technology Blanchardstown Dublin, Ireland.

[19] Chandola, V., S.R. Sukumar and J.C. Schryver. 2013. Knowledge discovery from massive healthcare claims data. pp. 1312–1320. *In*: Proceedings of the 19th ACM SIGKDD International Conference on Knowledge Discovery and Data Mining, ACM.

[20] Chawla, N.V. 2009. Data mining for imbalanced datasets: An overview. pp. 875–886. *In*: Data Mining and Knowledge Discovery Handbook. Springer.

[21] CMS. HCPCS—General Information. [Online]. Available: https://www.cms.gov/ Medicare/Coding/MedHCPCSGenInfo/index.html?redirect=/medhcpcsgeninfo/.

[22] CMS. Medicare provider utilization and payment data: Physician and other supplier. [Online]. Available: https://www.cms.gov/Research-Statistics-Data-and-Systems/ Statistics-Trends-and-Reports/Medicare-Provider-Charge-Data/Physician-and-Other-Supplier.html.

[23] CMS. National provider identifier standard (npi). [Online]. Available: https://www.cms.gov/Regulations-and-Guidance/Administrative-Simplification/NationalProvIdentStand/.

[24] CMS. 2016. What's medicare. [Online]. Available: https://www.medicare.gov/sign-up-change-plans/decide-how-toget-medicare/whats-medicare/what-is-medicare.html.

[25] CMS. 2017. Medicare fraud & abuse: Prevention, detection, and reporting booklet. [Online]. Available: https://www.cms.gov/Outreach-and-Education/Medicare-Learning-Network-MLN/MLNProducts/downloads/fraud_and_abuse.pdf.

[26] CMS Office of Enterprise Data and Analytics. 2017. Medicare fee-for-service provider utilization & payment data physician and other supplier. [Online]. Available: https://www.cms.gov/Research-Statistics-Data-and-Systems/Statistics-Trends-and-Reports/Medicare-Provider-Charge-Data/Downloads/Medicare-Physician-and-Other-Supplier-PUF-Methodology.pdf.

[27] CMS Office of Enterprise Data and Analytics. 2018. Medicare fee-for-service provider utilization & payment data part D prescriber public use file: A methodological overview. [Online]. Available: https://www.cms.gov/Research-Statistics-Data-and-Systems/Statistics-Trends-and-Reports/Medicare-Provider-Charge-Data/Downloads/Prescriber_Methods.pdf.

[28] CMS Office of Enterprise Data and Analytics. 2018. Medicare fee-for-service provider utilization & payment data physician and other supplier. [Online]. Available: https://www.cms.gov/Research-Statistics-Data-and-Systems/Statistics-Trends-and-Reports/Medicare-Provider-Charge-Data/Downloads/Medicare-Physician-and-Other-Supplier-PUF-Methodology.pdf.

[29] CMS Office of Enterprise Data and Analytics. 2018. Medicare fee-for-service provider utilization & payment data referring durable medical equipment, prosthetics, orthotics and supplies public use file: A methodological overview. [Online]. Available: https://www.cms.gov/Research-Statistics-Data-and-Systems/Statistics-Trends-and-Reports/Medicare-Provider-Charge-Data/Downloads/DME_Methodology.pdf.

[30] CMS Outreach and Education. 2017, Sep. Medicare Fraud & abuse: prevention, detection, and reporting. [Online]. Available: https://www.cms.gov/Outreach-and-Education/Medicare-Learning-Network-MLN/MLNProducts/downloads/fraud_and_abuse.pdf.

[31] Coalition Against Insurance Fraud. 2018. By the numbers: fraud statistics. [Online]. Available: http://www.insurancefraud.org/statistics.htm.

[32] Cubanski, J. and T. Neuman. 2018. The facts on medicare spending and financing. Henry J. Kaiser Family Foundation. [Online]. Available: https://www.kff.org/medicare/issue-brief/the-facts-on-medicare-spending-and-financing/.

[33] Feldstein, M. 2006. Balancing the goals of health care provision and financing. Health Affairs 25(6): 1603–1611.

[34] Fernández-Delgado, M., E. Cernadas, S. Barro and D. Amorim. 2014. Do we need hundreds of classifiers to solve real world classification problems. J. Mach. Learn. Res. 15(1): 3133–3181.

[35] Gelman, A. et al. 2005. Analysis of variance: why it is more important than ever. The Annals of Statistics 33(1): 1–53.

[36] Hasanin, T. and T.M. Khoshgoftaar. 2018. The effects of random undersampling with simulated class imbalance for big data. pp. 70–79. *In*: IEEE International Conference on Information Reuse and Integration (IRI), IEEE.

[37] Herland, M., R.A. Bauder and T.M. Khoshgoftaar. 2019. The effects of class rarity on the evaluation of supervised healthcare fraud detection models. Journal of Big Data 6(1): 21.

[38] Herland, M., T.M. Khoshgoftaar and R. Wald. 2014. A review of data mining using big data in health informatics. Journal of Big Data 1(1): 2.

[39] Jeni, L.A., J.F. Cohn and F. De La Torre. 2013. Facing imbalanced data–recommendations for the use of performance metrics. pp. 245–251. *In*: Affective Computing and Intelligent Interaction (ACII), 2013 Humaine Association Conference on, IEEE.

[40] Khoshgoftaar, T.M., M. Golawala and J. Van Hulse. 2007. An empirical study of learning from imbalanced data using random forest. pp. 310–317. *In*: Tools with Artificial Intelligence, 2007. ICTAI 2007. 19th IEEE International Conference on, Vol. 2, IEEE.

[41] Khoshgoftaar, T.M., C. Seiffert, J. Van Hulse, A. Napolitano and A. Folleco. 2007. Learning with limited minority class data. pp. 348–353. *In*: Machine Learning and Applications, 2007. ICMLA 2007. Sixth International Conference on, IEEE.

[42] Khurjekar, N., C.-A. Chou and M.T. Khasawneh. 2015. Detection of fraudulent claims using hierarchical cluster analysis. *In*: IIE Annual Conference. Proceedings. Institute of Industrial and Systems Engineers (IISE), p. 2388.

[43] Ko, J.S., H. Chalfin, B.J. Trock, Z. Feng, E. Humphreys, S.-W. Park, H.B. Carter, K.D. Frick and M. Han. 2015. Variability in medicare utilization and payment among urologists. Urology 85(5): 1045–1051.

[44] Lazer, D., R. Kennedy, G. King and A. Vespignani. 2014. The parable of google flu: traps in big data analysis. Science 343(6176): 1203–1205.

[45] Leevy, J.L., T.M. Khoshgoftaar, R.A. Bauder and N. Seliya. 2018. A survey on addressing high-class imbalance in big data. Journal of Big Data 5(1): 42.

[46] LEIE. 2017. Office of inspector general leie downloadable databases. [Online]. Available: https://oig.hhs.gov/exclusions/index.asp.

[47] Marr, Bernard. 2015, Sep. 4 Ways Big Data Will Change Every Business. [Online]. Available: https://www.forbes.com/sites/bernardmarr/2015/09/08/4-ways-big-data-will-change-every-business/.

[48] Marr, Bernard. 2015, Apr. How Big Data Is Changing Healthcare. [Online]. Available: https://www.forbes.com/sites/bernardmarr/2015/04/21/how-big-data-is-changing-healthcare/#1345d00a2873.

[49] Morris, L. 2009. Combating fraud in health care: An essential component of any cost containment strategy. [Online]. Available: https://www.healthaffairs.org/doi/abs/10.1377/hlthaff.28.5.1351.

[50] Pande, V. and W. Maas. 2013. Physician medicare fraud: characteristics and consequences. International Journal of Pharmaceutical and Healthcare Marketing 7(1): 8–33.

[51] Prati, R.C., G.E. Batista and D.F. Silva. 2015. Class imbalance revisited: a new experimental setup to assess the performance of treatment methods. Knowledge and Information Systems 45(1): 247–270.

[52] Rashidian, A., H. Joudaki and T. Vian. 2012. No evidence of the effect of the interventions to combat health care fraud and abuse: A systematic review of literature. PloS One 7(8): e41988.

[53] Roesems-Kerremans, G. 2016. Big data in healthcare. Journal of Healthcare Communications.

[54] Sadiq, S., Y. Tao, Y. Yan and M.-L. Shyu. 2017. Mining anomalies in medicare big data using patient rule induction method. pp. 185–192. *In*: Multimedia Big Data (BigMM), 2017 IEEE Third International Conference on, IEEE.

[55] Schulte, F. 2017. Fraud and billing mistakes cost medicare—and taxpayers—tens of billions last year. Henry J. Kaiser Family Foundation. [Online]. Available: https://khn.org/news/fraud-and-billing-mistakes-cost-medicare-and-taxpayers-tens-of-billions-last-year/.

[56] Seiffert, C., T.M. Khoshgoftaar, J. Van Hulse and A. Napolitano. 2007. Mining data with rare events: a case study. pp. 132–139. *In*: Tools with Artificial Intelligence, 2007. ICTAI 2007. 19th IEEE International Conference on, Vol. 2, IEEE.

[57] Seliya, N., T.M. Khoshgoftaar and J. Van Hulse. 2009. A study on the relationships of classifier performance metrics. pp. 59–66. *In*: Tools with Artificial Intelligence, 2009. ICTAI'09. 21st International Conference on, IEEE.

[58] Tukey, J.W. 1949. Comparing individual means in the analysis of variance. Biometrics 5(2): 99–114. [Online]. Available: http://www.jstor.org/stable/3001913.

[59] Van Hulse, J., T.M. Khoshgoftaar and A. Napolitano. 2007. Experimental perspectives on learning from imbalanced data. pp. 935–942. *In*: Proceedings of the 24th International Conference on Machine Learning, ACM.

[60] Wang, S.-L., H.-T. Pai, M.-F. Wu, F. Wu and C.-L. Li. 2017. The evaluation of trustworthiness to identify health insurance fraud in dentistry. Artificial Intelligence in Medicine 75: 40–50. [Online]. Available: http://www.sciencedirect.com/science/article/pii/S0933365716300513.

[61] Witten, I.H., E. Frank, M.A. Hall and C.J. Pal. 2016. Data Mining: Practical Machine Learning Tools and Techniques. Morgan Kaufmann, 51.

Chapter 4

Movie Recommendations Based on a Recurrent Neural Network Model

Yiu-Kai Ng

1. Introduction

Movie streaming services like Netflix, Hulu, Amazon Prime, and others are increasingly used by consumers to discover video content. For example, in 2017 Netflix subscribers collectively watched more than 140 million hours per day[1] and Netflix surpassed $11 billion in revenue in 2017.[2] In fact, roughly 80% of hours streamed at Netflix were influenced by their proprietary recommendation system [12]. Undoubtedly, movie streaming services have become an integral part of how we consume video content today, and the importance of movie recommendation systems cannot be understated—they are an integral part of how we consume video content today. With this in mind, the problem we propose to work on is movie recommendations through collaborative filtering based on the deep learning strategy.

For movie streaming services like Netflix, recommendation systems are important for helping users to discover new content to enjoy. While the

---

Computer Science Department, Brigham Young University, Provo, Utah 84602, USA.
  Email: ng@compsci.byu.edu

[1] techcrunch.com/2017/12/11/netflix-users-collectively-watched-1-billion-hours-of-content-per-week-in-2017/.

[2] tvtechnology.com/news/netflix-surpasses-11-billion-in-2017-revenue.

details of this system are mostly confidential, what we do know is that it is a combination of various individual recommendation systems, including some systems which leverage collaborative filtering systems [15]. In light of this, the problem we examine is movie recommendations through collaborative filtering.

Collaborative filtering is an approach for recommendation systems which relies on the ratings for a particular user as well as the ratings of similar users. The underlying assumption is that if we can accurately predict movie ratings, then we can recommend new movies to users that they are likely to enjoy, including movies the user may not have considered before. Therefore, in the context of movie recommendation, collaborative filtering aims to predict unknown movie ratings for a particular user, based on that user's known ratings as well as the movie ratings by other users in the system. As opposed to content-based systems, collaborative filtering accounts for users with diverse taste, so long as there are other users with similar preferences. By finding similar users, new items can be recommended based on the assumption that items which are liked by similar users will be liked by the user in question.

There are many ways to perform collaborative filtering such as utilizing k-nearest neighbor clustering with user profiles [6]. Various approaches for measuring similarity have been proposed, but a simple approach is to represent a user profile as a vector, and then use some measure of similarity between those vectors (e.g., cosine similarity). An alternative k-nearest-neighbor approach instead computes similarity between pairs of items with the idea that users who like a particular item will like similar items [28]. Another common method for performing collaborative filtering is with matrix factorization [18]. With this technique a user-item matrix is factorized into two matrices with the inner dimension representing some latent factors. The resulting factorization represents both users and items in terms of the latent factors in such a way that new items can be recommended to users based on the latent factors.

Lately, deep learning has demonstrated its effectiveness in coping with recommendation tasks. Due to its state-of-the-art performances and high-quality recommendations, deep learning techniques have been gaining momentum in recommender system. Compared with traditional recommendation models, deep learning provides a better understanding of user's demands, item's characteristics, and historical interactions between them. We apply the deep learning approach for movie recommendation.

The rest of the paper is organized as follows. The most popular approaches for collaborative filtering are discussed in Section 2. These methods work by computing neighborhoods of similar users or items. In contrast, in Section 3 we propose a deep learning approach for collaborative filtering based on an autoencoder. We demonstrate in Section 4 that our approach outperforms the neighborhood-based baseline. We give a concluding remark in Section 5.

2. Related Work

The most common method of performing collaborative filtering is to utilize a k-nearest-neighbor approach between users [6]. With this technique, it first starts with a user-item matrix R, where $R_{i,j}$ gives the rating of user i for item j and the value 0 indicates that a particular rating is missing. From R a user-user similarity matrix S is computed, where $S_{i,j}$ is the similarity between user i and user j, which can be computed with $R \cdot R^T$. Note that using other distance metrics, such as the correlation similarity measure or cosine similarity, to populate S are also effective. Once S is computed, we can predict the rating of user i for item j by computing $R_j^T \cdot S_i$, which essentially computes the average of the other users' ratings for item j weighted by their similarity to user i.

We can also use the k most similar users to user i to predict the rating for item j. Empirically, this works better than the weighted average over all users, although some extra work is required at test time in order to compute the k nearest neighbors. This approach relies on the assumption that if two users rated the same item similarly, they are likely to rate other items similarly as well. At scale, data structures such as ball trees [21] and k-d trees (a binary space partition tree in k-dimensions) can be utilized to more efficiently compute local neighbors between user profiles.

An alternative k-nearest-neighbor approach instead computes similarity between pairs of items (as opposed to users) with the idea that users who like a particular item will like similar items [28]. With this approach we compute an item-to-item similarity matrix I as $R^T \cdot R$. As before, we can also use other similarly metrics to populate I. In order to predict the rating for user i on item j, we can compute $R_i \cdot I_j$, which gives an average of the ratings provided by user i weighted by the similarity of those items to item j. Since there tends to be many more users than items in a recommender system, user-user collaborative filtering can be more performant, although our preliminary experiments with movie ratings indicate that user-user produced more accurate predictions of movie ratings.

Another common method for performing collaborative filtering is with matrix factorization [18]. With this technique a user-item matrix is factorized into two matrices with the inner dimension representing some latent factors using techniques such as singular value decomposition (SVD) [19]. The resulting factorization represents both users and items in terms of the latent factors in such a way that they can be used to recommend new items. As with item-item neighborhood approaches, our preliminary experiments on movie ratings indicate that user-user neighborhood approaches are superior to matrix factorization.

Deep learning has revolutionized many fields of computer science, including natural language processing [22]. Despite this, deep learning is relatively new in the area of recommender systems, and has not received much attention [40]. Having said that, Wang et al. [35] propose a collaborative deep learning (CDL) model which jointly performs deep representation learning for the content information and collaborative filtering for the rating matrix. CDL is differed from ours, since the former relies on content information, whereas we do not. Elkahky et al. [8] introduce a deep learning recommendation system according to the web browsing history and search queries provided by users. They maximize the similarity between users and their preferred items by mapping users and items to a latent space. A constraint imposed on this approach is that browsing history and users search queries are required, which are not always available. Wei et al. [36] develop a deep neural network model which extracts the content features of items into prediction of ratings for cold start items, which again is differed from ours, since we do not deal with user content.

Deep Learning provides a new toolkit for recommender systems designers and developers to extract features and to model user generated data and item data that has the potential to provide large improvements in the quality of the recommendations provided to users [16]. Part of the power of deep learning techniques in recommender systems stems from the fact that deep learning methods allow for much better feature extraction from item characteristics such as image, video, and audio compared to traditional techniques. Our recommender system, which is based on Recurrent Neural Networks, uses the autoencoder network directly on the user item interactions in order to build collaborative filtering models that can then be used for recommendations. This method can be treated as a form of deep factorization methods, which often outperform standard model-based collaborative filtering methods [38]. The aim of this paper is to experiment with deep learning for collaborative filtering on a large set of movie ratings.

3. Our Proposed Recommendation System

Deep learning, which is essentially just deep artificial neural networks, is able to learn complex decision boundaries for classification or complex non-linear regressions. By stacking large numbers of hidden layers in these networks, deep neural networks can learn complex functions by learning to extract many low level features from the data and composing them in useful non-linear combinations. Figure 1 depicts the system architecture of our deep learning model.

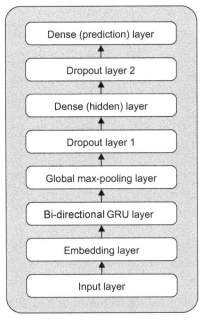

Fig. 1: The system architecture of our deep learning model.

3.1 The Recurrent Neural Network (RNN) Model

We employ a recurrent neural network (RNN) as our classifier for predicting movies ratings, with initial ratings on movies provided by users to begin with, since RNNs have been proven to produce robust models for rating prediction [27]. A RNN is similar to other deep neural networks (DNNs) [20] in that they are both trained (optimized) by the backpropagation of errors [26] and are comprised of a series of layers. Table 1 summaries different layers, their dimensions, and their parameters in our RNN, in which 10 in the output dimensions (of the Dense Output layer) denotes the different rating values (from 0.5 to 5, with an incremental value of 0.5) predicted by the model.

The output is produced by propagating numeric values forward. The network is trained by backpropagating the *error*[3] from the output layer backwards. Unlike other network structures, a RNN takes into account the *ordering* of tokens within sequences, rather than simply accounting for the existence of certain values or combinations of values in that sequence.

While neural networks are theoretically able to approximate any computable function, including the mapping from user profiles to movie

---

[3] An error is the relative divergence of the produced output from the ground truth.

Table 1: Dimensions and number of parameters of layers in the RNN.

Layer	Output Dimensions	Total Parameters	Trainable Parameters
Input	72	0	0
Embedding	72 × 300	1,950,000	0
Bi-directional GRU	72 × 128	140,160	140,160
Global Max Pooling, 1D	128	0	0
Dropout 1	128	0	0
Dense Hidden	64	8,256	8,256
Dropout 2	64	0	0
Dense Output	10	650	650
Total		**2,099,066**	**149,066**

ratings, in practice great care must be taken when selecting the architecture of the neural network. While the extracted structure of our network is subject to change, there are some reasonable starting places.

Inputs. The inputs to our network architecture are two n-dimensional vectors, where n is the number of movies in a movie dataset, such as the MovieLens database. One vector encodes a particular user profile, with each dimension indicating the rating the user gave for a particular film (or a zero to indicate that no rating has been given). The other vector is a one-hot encoding of a particular movie (i.e., a vector with a single "hot" dimension set to 1, with all other values set to zero). These two vectors request that the network predict a rating for a particular user for a specific movie.

One advantage of this input format is that we can do without a single rating from a known user profile, and use the known rating for withheld item as a labeled example. Consequently, even though we only have 270,000 user profiles created by using the MovieLens dataset, each one of the 26,000,000 individual ratings constitutes a train example.

Hidden Layers. There are a variety of ways to structure a simple feed-forward neural network. We start with a number of the standard fully-connected layers. However, we also experiment with alternative structures, such as ResNets [14], which currently obtain state-of-the-art results in other fields such as image recognition.

Output. There are two main possibilities for the output of our network. The first is to treat this problem as a classification problem, with ten different class representing the ten start ratings that are present in the data. Under this architecture, we treat the ten outputs of our network as unnormalized log probabilities, and use cross entropy as our loss function.

RNNs achieve the recurrent pattern matching through its *recurrent layer(s)*. A recurrent layer is one which contains a single recurrent unit through which each value of the input vector or matrix passes. The recurrent unit maintains a *state* which can be thought of as a "memory". As each value in the input iteratively passes through the unit at time step t, the unit updates its state h_t based on a function of that input value x_t and its own previous state h_{t-1} as

$$h_t = f(h_{t-1}, x_t) \tag{1}$$

where f is any non-linear activation.

Throughout training, the unit learns, i.e., optimizes, this state-updating function—it learns how much of its current state to keep or discard as it processes certain input values. Although the layer contains just a single unit, it can be visualized to have a number of units equal to its number of time steps, or iterations of processing sequential input values and previous states. This architecture is shown in Figure 2 [7].

Recurrent layers are designed to "remember" the most important features in sequenced data no matter if the feature appears towards the beginning of the sequence or the end. In fact, one widely-used implementation of a recurrent unit is thus named "Long-Short Term Memory", or LSTM [10]. RNNs have been shown to be effective tools in fields such as language modeling and speech recognition. The designed RNN accurately predicts movie ratings solely based on the sequential of given user ratings.

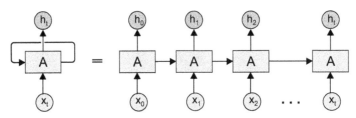

Fig. 2: The actual structure of a recurrent layer (left), and an "unrolled" representation of the recurrent layer through t time steps (right).

3.1.1 Feature Representation

In order to utilize a RNN, we need to provide the network with sequential data as input and a corresponding ground-truth value as its target output. Each of the data entries has to first been transformed in order to be fed into the RNN. Attributes of movie ratings are manipulated as labels, which are the naming of the categories of movie ratings, which are the categories pre-defined from 0.5 to 5, with a half-star interval. Since neural networks cannot accept strings as an output target, each unique category string is assigned a unique integer value, which is transformed into a one-hot encoding to be used later as the network's prediction target. A one-hot encoding of an integer value i among n unique values is a binarized representation of that integer as an n-dimensional vector of all zeros except the ith element, which is a one. For example, if a movie rating is assigned the value 4, then with 10 distinct labels, its one-hot encoding is [0 0 0 0 0 0 0 1 0 0].

3.1.2 Network Structure

In this section, we explain the technical details of the RNN used for predicting movie ratings.

The Embedding Layer. One of the design goals of our neural network is to capture relatedness between similar user ratings for different movies. Due to the large amount of time it would take to properly train the embedding from scratch, we have performed two different tasks: (i) we have loaded into the embedding layer as weights an uncased ratings, GloVe [23], which has been pre-trained on movie ratings extracted from the MovieLens dataset, and (ii) we have decided to freeze, i.e., not train, the embedding layer at all. The pretrained vectors from GloVe sufficiently capture different ratings for our task and they are not required to be further optimized.

The Bi-directional GRU Layer. Following the embedding layer in our network is one type of recurrent layer—a bi-directional GRU, or Gated Recurrent Unit [5], layer. Figure 3 shows the architecture of a GRU layer [2].

A GRU is a current state-of-the-art recurrent unit which is able to 'remember' important patterns within sequences and 'forget' the unimportant ones. The original architecture of a gated recurrent unit proposed by Cho

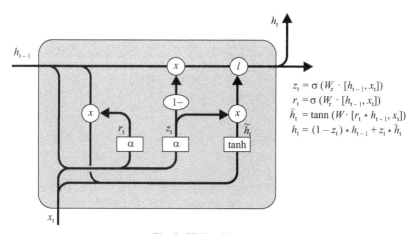

$$z_t = \sigma \left(W_z \cdot [h_{t-1}, x_t] \right)$$
$$r_t = \sigma \left(W_r \cdot [h_{t-1}, x_t] \right)$$
$$\tilde{h}_t = \tanh \left(W \cdot [r_t * h_{t-1}, x_t] \right)$$
$$h_t = (1 - z_t) * h_{t-1} + z_t * \tilde{h}_t$$

Fig. 3: GRU architecture.

et al. [4]—and the one which we used for our task—computes each subsequent hidden state h_t as a function of its previous state and current inputs (as defined in Equation 1) as follows:

- Input and previous state values pass through two "gates", or intermediate value stages, before the final hidden state h_t is computed. First, the *reset* gate r_t is computed as

$$r_t = \sigma(W_r \cdot [h_{t-1}, x_t] + b_r) \tag{2}$$

where $\sigma(x) = \frac{1}{1 + e^{-x}}$, which is the logistic sigmoid function in the range between 0 and 1, $[.]_t$ denotes the tth element in a vector, x_t and h_{t-1} are the current input and the previous hidden state, respectively, W_r is a learned weight matrix, and b_r is a bias vector.

- The *update* gate z_t is similarly computed as

$$z_t = \sigma(W_z \cdot [h_{t-1}, x_t] + b_z) \tag{3}$$

where W_z is another learned weight matrix and b_z is another bias vector. A candidate hidden state, $\tilde{h}_t$, is then computed as

$$\tilde{h}_t = tanh(W_h \cdot [r_t \times h_{t-1}, x_t] + b_h) \tag{4}$$

where *tanh* is the hyperbolic tangent function and W_h and b_h are another learned weight matrix and bias vector, respectively.

- The *hidden state*, h_t, is produced as

$$h_t = z_t \times \tilde{h}_t + (1 - z_t) \times h_{t-1} \tag{5}$$

The value of z_t in Equation 5 guides the unit's decision of whether to update the hidden state (when z_t is close to 1) or to leave it mostly unchanged (when z_t is close to 0).

The number of trainable parameters in a single GRU layer is $3 \times (n^2 + n(m + 1))$, where n is the output dimension, or the number of time steps through which the input values pass, and m is the input dimension. In our case, $n = 64$, since we have chosen to pass each input through 64 time steps, and $m = 300$. Since our layer is bi-directional, the number of trainable parameters is twice that of a single layer, i.e., $2 \times 3 \times (64^2 + 64 \times 301) = 140$, 160, the greatest number of trainable parameters in our network as shown in Table 1.

The recurrent layer outputs a 72×128 matrix, where 72 represents the number of tokens in a sequence, and 128 denotes the respective output values of the GRU after each of 64 time steps in two directions.

The Global Max-Pooling Layer (1D). At this point in the network, it is necessary to reduce the matrix output from the GRU layer to a more manageable vector which we eventually use to classify the token sequence into one of the movie rating categories. In order to reduce the dimensionality of the output, we pass the matrix through a *global max-pooling* layer. This layer simply returns as output the maximum value of each column in the matrix. Max-pooling is one of several pooling functions, besides sum- or average-pooling, used to reduce the dimensionality of its input. Pooling can be done in more than one direction. For example, the image detection systems commonly pool a 2-dimensional area of an image into a scalar value. Since pooling is a computable function, not a learnable one, this layer cannot be optimized and contains no trainable parameters. The output of the max-pooling layer is a 128-dimensional vector.

The Dropout Layer 1. Our model includes at this point a dropout layer [9]. Dropout, a common technique used in deep neural networks which helps to prevent a model from overfitting, occurs when the output of a percentage of nodes in a layer are suppressed. (See Figure 4 for an example of a dropout layer [32].) The nodes which are chosen to be dropped out are probabilistically determined at each pass of data through the network. Since dropout does not change the dimensions of the input, this layer in our network also outputs a 128-dimensional vector.

The Dense Hidden Layer. Our RNN model includes a dense, or fully-connected, layer as shown in Figure 5 [25]. A *dense layer* is typical of nearly all neural networks and is used for discovering hidden, or latent, features from the previous layers. It transforms a vector x with N elements into a vector y with M inputs by multiplying x by a $M \times N$ weight matrix W. Throughout training, weights are optimized via backpropagation.

The Dropout Layer 2. Before classification, our RNN model includes another dropout layer to again avoid overfitting to the training sequences.

The Dense Output Layer. At last, our RNN model includes a final dense layer which outputs ten distinct values, each value corresponding to the relative probability of the input belonging to one of the ten unique categories. Each instance is classified according to the category corresponding to the highest of the 10 output values.

Our bi-directional GRU layer also features dropout of each time step's output value and of its recurring state. Each dropout layer's probability of "dropping" its output values is set to 10%. Furthermore, the network is trained by the Rmsprop optimization algorithm [34], though many other optimizers have been shown to perform similarly.

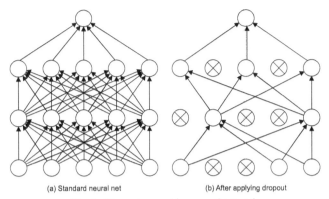

(a) Standard neural net   (b) After applying dropout

Fig. 4: Dropout as used in a neural network.

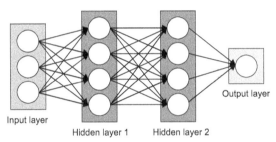

Input layer

Hidden layer 1   Hidden layer 2

Output layer

Fig. 5: Structure of a dense layer.

3.2 Network Architecture

With the neural network architecture introduced above, we describe the deep learning architecture proposed as an alternative to the user-based neighborhood approach. We first consider the dimensions of the input and output of the neural network. In order to maximize the amount of training data we can feed to the network, we consider a training example to be a user profile (i.e., a row from the user-item matrix R) with one rating withheld. The loss of the network on that training example must be computed with respect to the single withheld rating. The consequence of this is that each individual rating in the training set corresponds to a training example, rather than each user.

As we are interested in what is essentially a regression, we choose to use *root mean squared error (RMSE)* with respect to known ratings as our *loss function*. Compared to the mean absolute error, root mean squared error more heavily penalizes predictions which are further off. We reason that this is good in the context of recommender system because predicting a high rating for an item the user did not enjoy significantly impacts the quality of the recommendations. On the other hand, smaller errors in prediction likely result in recommendations that are still useful—perhaps the regression is not exactly correct, but at least the highest predicted rating are likely to be relevant to the user.

3.2.1 Autoencoder

One of the existing deep learning models is the Deep Neural Network (DNN) model. DNN is a Multi-Layer Perceptron (MLP) model with many hidden layers. The uniqueness of DNN is due to its larger number of hidden units and better parameter initialization techniques. A DNN model with large number of hidden units can have better modeling power. Although the learned parameters of the DNN model is a local optimal, which requires more training data and more computational power, it can perform much better than those with less hidden units. Deep Auto Encoder is a special type of DNN. (See Figure 6 for a sample autoencoder [3].)

An autoencoder is a neural network that is trained to copy its input to its output, with the typical purpose of dimension reduction, i.e., the process of reducing the number of random variables under consideration. It features an encoder function to create a hidden layer (or multiple hidden layers) which contains a code to describe the input. There is a decoder which creates a reconstruction of the input from the hidden layer. An autoencoder can then become useful by having a hidden layer smaller than the input layer, forcing it to create a compressed representation of the data in the hidden layer by learning correlations in the data. This autoencoder is a form of unsupervised learning, meaning that an autoencoder only needs unlabelled data, which is a set of input data rather than input-output pairs. Through an unsupervised learning algorithm, for linear reconstructions the autoencoder attempts to learn a function to minimize the root mean square difference.

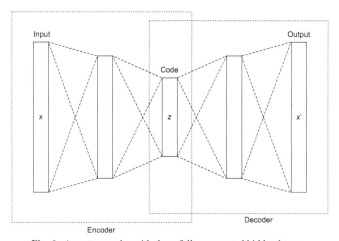

Fig. 6: An autoencoder with three fully-connected hidden layers.

To compute the *root mean square error* (RMSE) of a machine learning model, we can measure the performance of the model. RMSE is defined as

$$RMSE = \frac{1}{m}\sum_{i}(\hat{y}-y)_i^2, \qquad \hat{y} = w^T x \qquad (6)$$

where $w \in \mathfrak{R}^n$ is a vector of parameters, $x \in \mathfrak{R}^n$ is a vector used for predicting a scalar value $y \in \mathfrak{R}$, and $\hat{y}$ is the value that a machine learning model predicts what the scalar value $y \in \mathfrak{R}$ should be.

Note that RMSE decreases to 0 when $\hat{y} = y$ and the error increases when the *Euclidean distance* between the predicted values and the target values increases.

3.2.2 Multilayer Perceptron

Initially, the architecture of our recommender system consists of input from the row of the user-item matrix R with the rating for some item j withheld, along with a one-hot encoded query which indicates the network should predict the rating for user i on item j. Unfortunately, this architecture has been proved difficult to train, since the network must learn to understand not only user profiles, but also the interplay between those profiles and the query inputs. With respect to the root mean squared error on the training data, we never achieved a loss less than 1.2 with this architecture.

Instead, we take inspiration from the concept of an *autoencoder* to design our neural network architecture. This simple architecture takes an input and connects it to some number of fully connected hidden layers which include a "bottleneck." This bottleneck is a hidden layer which has a much smaller dimensionality than the input. The output of the network is then re-expanded to have the same dimensionality as the input. The network is then trained to learn the identity function, with the idea that in order for the network to compute the identity function through the bottleneck, it must learn a dense representation of the input. Thus, the autoencoder could be viewed as something akin to a *dimensionality reduction* technique. We can also hope that the bottleneck layer learns something useful related to the structure underlying the input. For example, a neuron in the bottleneck layer might represent something related to the genre of a movie or similar movie groupings.

Note that we are not interested in learning to compute an identity function—after all, our goal is to predict missing ratings, not reproduce the zeros in the input vectors. Consequently, while our final network architecture resembles an autoencoder with the bottleneck hidden layers and the matching dimensions on input and output, the network is actually trained using a loss function for regression (i.e., RMSE) with the aim of learning to predict missing ratings.

More specifically, the training examples to the network are user profiles with one rating withheld, and the output is the predicted ratings for all movies in the dataset. While the network is expected to predict ratings for each movie based on a user profile, we only have the answer for the one withheld rating. Consequently, we only propagate loss for the missing rating when learning from the training example.[4]

3.2.3 *The Deep Learning Recommender System*

Withholding ratings does have the unfortunate consequence that our deep learning model is only able to learn ratings for movies similar to what the user has actually watched, as the loss function is not directly affected by the output on unrelated movies. Due to the bottleneck layer, the model is required to generalize to some degree, but the model may have difficulty for movies which are drastically different than the movies the user actually rated. While users do watch movies they rate lowly, most of the time they do not rate more than a few hundred items, and avoid watching completely non-relevant movies, so it may be difficult for the model to predict ratings for completely unrelated movies.

For the purposes of our loss function, which is *root mean squared error* on known ratings, the fact that our network may not learn how to output ratings for completely unrelated movies does not seem to affect the test loss, probably because the movies in the test data are related enough that the patterns learned from the training data generalize to the ratings in the test data. Of course, it may affect the rankings, so it could be desirable to add a *regularization* term (discussed in details in Section 3.3) to the loss, which encourages sparsity in the output.

With this basic design in place, we have experimented with several variations of this architecture using various numbers of layers, and various sizes for the bottleneck layer. The most interesting parameter was the size of the smallest bottleneck layer, and after experimenting with various values, we eventually settled on a bottleneck size of 512. From there we experimented with different numbers of fully connected layers, always using powers of 2 to increase and decrease the dimensionality. The final network topology has seven fully connected hidden layers with dimensions [4096, 2048, 1024, 512, 1024, 2048, 4096]. Each layer used a rectified linear unit[5] as the non-linear activation function. The connecting weights of the hidden layers were initialized using Xavier initialization [11] with the biases set to zeros.

---

[4] In code, this can be accomplished with the tf.gather function.
[5] The rectified linear unit, or ReLU, is defined as $max(0; x)$. While simple, it is currently the state-of-the-art in activation functions for DNN.

3.2.4 Clustering

We have considered the idea of using the smallest bottleneck layer in the network as some form of a natural clustering. By forcing the input into such a small dimensional space, the model must necessarily learn something about the underlying structure of the input data. The hypothesis was that by fixing a single neuron in the bottleneck layer and zeroing out the remaining neurons in the bottleneck layer, and then optimizing the input space for this particular activation, we can visualize that structure by showing the movies which trigger each *cluster*. For example, we expect that there might be a neuron or small set of neurons which trigger for various genres of movie, or various styles of filmography.

Table 2 gives an example of such a "cluster" from optimizing the input to trigger a single bottleneck neuron. These movies have common theme. Obviously, for this network to be able to accurately predict movie ratings it must learn some sort of structure. However, this structure is more distributed throughout the bottleneck layer than expected. One potential solution to this problem is to add a regularization term to the loss which encourages sparsity in the bottleneck layer.

Table 2: A cluster when optimizing the input to trigger a single bottleneck neuron.

Jules and Jim (Jules et Jim) (1961)
Frankenstein Must Be Destroyed (1969)
Lolita (1962)
Lawnmower Man, The (1992)
First Knight (1995)
Urban Legends: Final Cut (2000)
Fair Game (1995)
Guinevere (1999)
Paradine Case, The (1947)
400 Blows The (Les quatre cents coups) (1959)

3.3 Regularization

Regularization in deep learning, and in machine learning in general, is an important concept which solves the overfitting problem. It is very important to implement the regularization while training a good model, since it is a technique used in an attempt to solve the overfitting problem.

As mentioned earlier, regularization is an attempt to correct for model overfitting by introducing additional information to the cost function. Within the context of least squares linear regression, the regularization term is added to a standard least squares linear regression cost function J as defined below.

$$J(\Theta) = \frac{1}{2} m [\sum_{i=1}^{m} (h_\Theta(x^i) - y^i)^2 + \lambda \sum_{j=1}^{n} \Theta_j^2] \tag{7}$$

where Θ is the parameter values, m is the number of training examples with n different features, $h_\Theta(x^i)$ is the estimator h_Θ value for the training example i, y^i is the actual labeled value of training example i, and λ is the *regularization constant*.

In discussing regularization we have employed L2 regularization, whereas L1 regularization is another such strategy for controlling overfitting. The two regularizations share the same goal but differ in a few key respects. Note that in Equation 7,

$$\lambda \sum_{j=1}^{n} \Theta_j^2 \tag{8}$$

is the L2 regularization term, whereas in L1, the same regularization term is written as

$$\lambda \sum_{j=1}^{n} |\Theta_j| \tag{9}$$

Hence, the difference between L1 and L2 is that L2 uses the sum of the square of the parameters, whereas L1 is the sum of the absolute value of the parameters. In essence, L1 regularization reduces some parameters associated with a given feature to zero, whereas L2 regularization does not set feature parameters to zero, but will only continue to reduce the value of a given Θ.

4. Experimental Results

In order to verify the performance of the proposed deep learning model in predicting the movie ratings accurately for movie goers so that they would enjoy the movies recommended by us, we have conducted various empirical studies, which compare the performance of our model with other state-of- the-art movie recommender systems. Prior to presenting the experimental results of our recommendation system, we discuss the dataset used for the empirical study and the experimental setup. We first describe the MovieLens dataset and then briefly explain the baseline model used as a point of comparison.

4.1 MovieLens Data

In academia the most well-known movie ratings dataset is undoubtedly the MovieLens dataset [13], although a close second is probably the Netflix prize

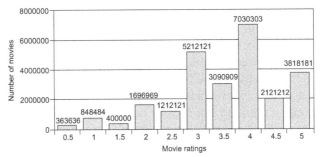

Fig. 7: Distribution of ratings in the full MovieLens dataset.

data released via Kaggle.[6] For our recommendation system we utilize the latest version of the MovieLens dataset, which is the recommended version for education and development.[7]

The MovieLens dataset is provided by GroupLens, which is a social computing research lab at the University of Minnesota. The full MovieLens dataset contains ratings for 45,115 movies provided by 270,896 different users. In total, the dataset contains 26,024,289 individual movie ratings, last updated in August 2017. Each rating allows users to assign between half a star and five stars to a movie, in half star increments. Figure 7 shows the distribution of the ratings in the data. Each rating is also accompanied by a time stamp. Since the dataset does not contain a standard train/test split, we used these time stamps to split the data into training and test sets, with the oldest 90% of the data making the training set and the newest 10% of the data composing the test set. We did this with the intent to mimic the problem faced by real world movie recommendation systems which have all of the data up to a certain point in time, and are faced with predicting movie ratings going forward in time.

4.2 Full Dataset Versus BaseLine

As previously mentioned, there are a number of popular methods for performing collaborative filtering, including nearest-neighbor based technique comparing user-user similarity [6], nearest-neighborhood comparing item-item similarity [28], and matrix factorization techniques [18]. We determined user-user neighborhood approach with cosine similarity and a neighborhood size of five performs the best with respect to root mean squared prediction error. In our empirical study, we used them all on the full MovieLens dataset. We allocated enough RAM to fully vectorize these algorithms. For example,

---

[6] https://www.kaggle.com/netfix-inc/netfix-prize-data.
[7] https://grouplens.org/datasets/movielens/latest.

in order to process the vectorized version of the user-user nearest neighbor approach, we computed a user-user similarity matrix which took nearly 600 GB in RAM. The non-vectorized brute force version of the algorithm required more than a week to finish. An alternative is to utilize a small version of the MovieLens dataset, called the BaseLine dataset, which contains only 943 users and 1,682 movies as a development dataset. The BaseLine database can be split into a train/test set, and we can measure the root mean squared error of the predictions of each of the proposed baseline algorithms.

4.3 Error Rates for Proposed Movie Recommenders

Using 90% of the full MovieLens dataset as training, we trained the architecture described in Section 3.2. It took roughly 4 days using a Titan X GPU to make 30 passes over the entire data before the training loss stabilized. Figure 8 shows the training loss (i.e., RMSE) decreasing over time.

We discuss the results of our model on the test set and compare its results to the user-based neighborhood models.

4.3.1 Root Mean Squared Error

Table 3 summarizes the results comparing our model-based approach with the user-based neighborhood baseline. On the training data, our approach is

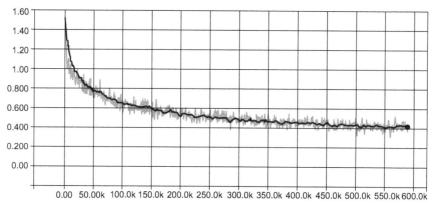

Fig. 8: Graph showing loss (root mean squared error) decreasing over time. Each step represents 1,000 training examples.

Table 3: Root mean squared error (RMSE) for our user-based neighborhood baseline and autoencoder inspired by our model-based approach.

	User-User KNN	**Model-based**
Train	N/A	0.4209
Test	11.6715	0.3544

stabilized around 0.42. The neighborhood approach has learned parameters, as it simply relies on the training data itself to make predictions. Consequently, there is no training loss to report.

On test data, our deep learning model-based algorithm outperforms the neighborhood approach by a large margin. However, it should be noted that for the purpose of making movie recommendations, we do not actually care about the error. Instead what we care about is the ranking of the top few most highly rated movies. It is not an unreasonable assumption that the algorithm which ranks *better* will also have *lower* root mean squared error, but it is entirely possible that despite the higher errors, the top ranked movies from the model-based approach produce superior recommendations. This is especially true when we consider that our algorithm does not directly learn about highly unrelated movies.

4.3.2 *Comparing our Movie Recommendation Systems with Others*

Besides using RMSE as shown in Table 3, we compare between various well-known movie recommenders and our deep learning movie recommendation model. These existing movie recommenders were chosen, since they achieve high accuracy in recommendations on movies based on their respective model, and more importantly they are simply based on user ratings, but not solely on contents.

- **MF.** Yu et al. [39] and Singh et al. [30] predict ratings on movies based on matrix factorization (MF), which can be adopted for solving large-scale collaborative filtering problems. Yu et al. develop a non-parametric matrix factorization (NPMF) method, which exploits data sparsity effectively and achieves predicted rankings on items comparable to or even superior than the performance of the state-of-the-art low-rank matrix factorization methods. Singh et al. introduce a collective matrix factorization (CMF) approach based on relational learning, which predicts user ratings on items based on the items' genres and role players, which are treated as unknown values of a relation between entities of a certain item using a given database of entities and observed relations among entities. Singh et al. propose different stochastic optimization methods to handle and work efficiently on large and sparse data sets with relational schemes. They have demonstrated that their model is practical to process relational domains with hundreds of thousands of entities.

- **ML.** Besides the matrix factorization methods, probabilistic frameworks have been introduced for rating predictions. Shi et al. [29] propose a joint matrix factorization model for making context-aware item recommendations. The matrix factorization model developed by Shi et al. relies not only on factorizing the user-item rating matrix but also considers contextual information of items. The model is capable of learning from user-item matrix, as in conventional collaborative filtering model, and simultaneously uses contextual information during the recommendation process. However, a significant difference between Shi et al.'s MF model and other MF approaches is that the contextual information of the former is based on movie mood, whereas other MF models makes recommendations according to the contextual information on movies.

- MudRecS [24], which makes recommendations on books, movies, music, and paintings similar in content to other books, movies, music, and/or paintings, respectively that a MudRecS user is interested in. MudRecS does not rely on users' access patterns/histories, connection information extracted from social networking sites, collaborated filtering methods, or user personal attributes (such as gender/age) to perform the recommendation task. It simply considers the users' ratings, genres, role players (authors or artists), and reviews of different multimedia items. MudRecS predicts the *ratings* of multimedia items that match the interests of a user to make recommendations.

- **Netflix.** We compare our deep learning recommendation system indirectly against the 20 systems that participated in the Netflix contest in 2008 through MudRecS [24]. The open competition was held by Netflix, an online DVD-rental service, and the Netflix Prize was awarded to the best recommendation algorithm with the lowest RMSE score in predicting user ratings on films based on previous ratings. On September 21, 2009, the grand prize of one million dollars were given. The RMSE scores achieved by each of the twenty systems, as well as detailed discussions on their rating prediction algorithms, can be found on the Netflix website.[8]

Figure 9 shows the Mean Absolute Error (MAE) and RMSE scores of our deep learning movie recommender and other recommendation systems on the MovieLens dataset. RMSE and MAE are two performance metrics widely-used for evaluating rating predictions on multimedia data [1]. Both RMSE and MAE measure the *average magnitude* of *error*, i.e., the average prediction error, on incorrectly assigned ratings. The error values computed by RMSE are squared before they are summed and averaged, which yield a relatively *high* weight to errors of *large* magnitude, whereas MAE is a *linear* score, i.e., the absolute values of individual differences in incorrect assignments are weighted equally in the average. Our deep learning recommender outperforms each of the movie recommenders as shown in Figure 9, and the RMSE and MAE values are statistically significant ($p < 0.01$) [31].

---

[8] https://www.netflixprize.com/leaderboard.html.

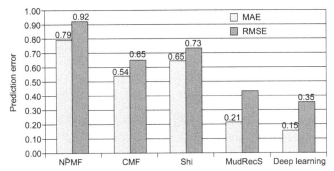

Fig. 9: The MAE and RMSE scores for various movie recommendation systems based on the MovieLens dataset.

On the *Netflix* dataset, MudRecS achieves a RMSE score[9] of 0.8571. MudRecS outperforms 18 recommendation systems and is only outperformed by two systems (Bellkor [17] and Ensemble [37]), both of which achieve the same score of 0.8567, a small, insignificant fraction (0.8571–0.8567 = 0.0004) better than MudRecS. The reason for the slightly better RMSE score achieved by the two systems on the Netflix dataset are twofold. Unlike MudRecS, Bellkor and Ensemble were specifically designed for movie rating predictions, and the construction of their algorithms focus on rating patterns found in movies which may not apply to other domains. Moreover, Bellkor and Ensemble account for temporal effects, i.e., the fact that a user's preference changes over time, which may lead to different ratings for the same movie over time. The temporal effect, however, does not apply to all users and requires a larger subset of training data in order to obtain reliable results, which are the constraints. In considering a 95% confidence interval, MudRecS significantly outperforms 17 recommendation systems and is *not* significantly outperformed by *any* of the twenty systems. CineMatch, Netflix's recommender, achieves an RMSE score of 0.9514 on the Netflix dataset, which is outperformed by MudRecS. We ran our deep learning recommender system on the Netflix dataset and achieves a 0.782 RMSE score, which is lower than MudRecS, even though the results are not statistically significant. However, our recommender performs at least as good as MudRecS based on the Netflix dataset.

4.4 Human Assessors

In order to further establish the usefulness of our deep learning approach in making movie recommendation, we conducted two user studies in which

---

[9] MAE scores were not computed on the Netflix dataset due to their unavailability for the other 20 recommenders.

users, who play the role of appraisers, had the chance to evaluate movies recommended by our system and the user-based neighborhood (KNN) approach in one case, and Amazon[10] and Redbox (www.redbox.com) in another.

4.4.1 College Student Appraisers

Appraisers were shown a user profile, which consisted of every movie the corresponding user had rated, as well as the associated ratings. Each appraiser was then presented two possible recommendations: one from our system and one from the user-based neighborhood approach. The recommendations were chosen by picking the movie with the highest predicted rating from either system, excluding movies that had already been rated by the user. The order in which the two possible recommendations were shown was randomized. Appraisers were asked to pick which recommendation they thought was more relevant to the given user profile (see Figure 10 for an example of the study).

A total of 100 participants, who were students at the authors' university, were used in the study. Each user, who is an appraiser, was asked to rate 15 randomly chosen recommendations. In this survey, 71.67% of the time appraisers preferred the recommendation made using our deep learning approach over the recommendation made by the baseline approach, and this result is encouraging. Of course, it is clear that this survey using a small sample size. In addition, most of the appraisers indicated that they were unfamiliar with most of movies referenced in the survey. Realizing this problem in advance, we indicated in the survey that they were allowed to use resources like Google[11] and IMDBa[12] while making their judgements.

Fig. 10: An example of the type of questions appraisers were asked to answer in the user evaluation of our deep learning-based system and the user-based KNN approach.

---

[10] www.amazon.com.
[11] https://www.google.com.
[12] www.imdb.com.

4.4.2 *Mechanical Turk Performance Evaluation*

Besides relying on college students to conduct a user study to evaluate the performed of our movie recommender *MR*, we also turned to Mechanical Turk[13] to conduct empirical studies that allow us to evaluate the performance of *MR*, which offer a diverse group of appraisers who come from all walks of life. We counted on Amazon's Mechanical Turk, since it is a "market-place for work that requires human intelligence", which allows individuals or businesses to programmatically access thousands of diverse, on-demand workers and has been used in the past to collect user feedback on various information retrieval/recommendation tasks. Altogether, we created a total of thirty-five HITs,[14] each of which consists of 10 tasks and each task includes 10 designated movies and their corresponding set of recommended movies. The Mechanical Turk appraisers who participated in the performance evaluation were asked to determine which one of the nine recommendations,[15] if there were any, were relevant movies with respective to the corresponding designated movie. The three movies marked as *relevant most often* by the appraisers were considered our *gold standard* for the designated movie (and the corresponding profile in the case of our recommender *MR*). Table 4 shows the top-3 recommendations suggested by *MR*, Amazon, and Redbox and the number of times each recommended movie was marked as relevant with respect to a designated movie by the corresponding appraisers.

Besides marking which recommendation was relevant, the appraisers were also asked (to the best of their knowledge) to order, i.e., rank, the recommendations in terms of their degrees of relevance with respect to the corresponding designated movie. Based on the gold standard on the relevance and rankings set by the appraisers, we determined whether the recommendations provided by *MR* and its competitors were truly relevant and the degree of accuracy of their corresponding rankings. Note that during the evaluation process, we *randomized* the order of the nine recommended movies and asked the appraisers to mark and rank the recommendations they believed to be relevant to the designated movie.

4.4.3 *Precision@K and MRR*

Users of a recommendation system tend to look at only the top part of the ranked result list to find relevant recommendations. Some search tasks have

---

[13] https://www.mturk.com/mturk/welcome.

[14] A Human Intelligence Task, or HIT, is a single, self-contained assignment that a Mechanical Turk appraiser works on.

[15] Three each from our recommender, Amazon, and Redbox, which were the top-3 recommendations made by the three recommendation systems, respectively. The appraisers had no idea which recommendation was made by which recommender.

Table 4: Top-3 recommendations for each of the five sample designated movies made by our Deep Learning model (denoted DLM), Amazon, and Redbox, and their respective frequency of relevance based on the gold standard established by the 20 appraisers.

Designated Movie	DLM			Amazon			Redbox		
	Rank 1	Rank 2	Rank 3	Rank 1	Rank 2	Rank 3	Rank 1	Rank 2	Rank 3
Inception	The Dark Knight	The Matrix	Interstellar	Skyscraper	Anchorman	Jurassic World: Fallen Kingdom	Mission of Honor	Never Grow Old	Chasing Bullitt
Relevant	11	16	12	7	15	7	6	9	8
Dance with Wolves	Unforgiven	Rain Man	Robin Hood: Prince of Thieves	Annapolis	The Last of the Mohicans	Legends of the Fall	Chimera Strain	For Love or Money	Mission of Honor
Relevant	12	10	9	13	15	7	8	6	7
Lord of the Rings: The Two Towers	Star Wars: the Empire Strikes Back	Gladiator	Inception	Pride & Prejudice	Son of God	A Knight's Tale	Mission of Honor	Never Grow Old	Chasing Bullitt
Relevant	4	19	12	7	4	12	10	7	14
Sound of Music	The Princess Bride	The Wizard of Oz	Snow White and the Seven Dwarfs	The Parent Trap	Pollyanna	The Aristocats	Fun with Slime: Part 2	Fund with Slime Part 1	Caption Morten and the Spider Queen
Relevant	17	16	11	17	6	18	9	6	6
Happy Feet	Shark Tale	Happy Feet Two	A Bug's Life	Monsters, Inc.	Sing	Trolls	Daphne and Velma	Monster Trucks	Cop and a Half New Recruit
Relevant	18	16	18	15	10	13	9	7	8
Avg. Relevance	12.6	15.4	12.4	11.8	9.6	11.4	8.4	7.0	8.6

only one relevant result (i.e., precision at rank 1, denoted P@1) in mind, i.e., the top-ranked recommendation is expected to be relevant and useful, whereas others consider the top-*n* ($2 \leq n \leq 10$) ranked recommendations. Since our recommender *MR* suggests up to three movies for each user's designated movie based on the profile of the user, we have evaluated the performance of *MR* based on P@1 and P@3, which is easy to compute, flexible to be averaged over the recommendations made for different designated movies to produce a single performance value, and is readily understandable.

After the gold standard for each one of the 350 test cases provided by the 20 appraisers were recorded, we calculated the metrics for the *average precision* at rank 1 (i.e., average P@1) and *average Precision* at rank 3 (i.e., average P@3). The average P@1 values measure the usefulness of the recommendations at rank 1, whereas average P@3 computes the ratio of the usefulness of the top-3 ranked recommendations. As shown in Figure 11, our recommender *MR* scored an average P@1 value of 0.71,[16] which is compared favorably with Amazon's 0.65 and Redbox's 0.38. In addition, *MR* scored an average P@3 value of 0.67, which is also more appealing than Amazon's 0.53 and Redbox's 0.37. All of these results are statistically significant based on the Wilcoxon Signed-Ranks Test ($p < 0.01$).

Besides measuring the usefulness of the top-ranked recommendations made by our recommender *MR*, we have also evaluated the performance of *MR* based on the evaluation metric MRR. MRR calculates the average of the reciprocal ranks at which the *first* useful recommendation (among all the ranked recommendations) for each designated movie based is made. The

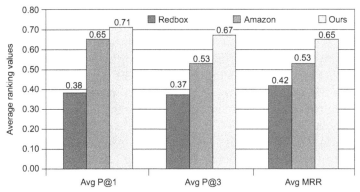

Fig. 11: Performance evaluation on average P@1, P@3, and MRR for Redbox, Amazon, and our movie recommender.

---

[16] The value 0.71 indicates that in seven out of 10 times the *first* recommendation made by *MR* is considered useful and relevant.

reciprocal rank is very sensitive to the rank position. MRR is formally defined below.

$$MRR = \frac{1}{|DGs|} \sum_{i=1}^{|DGs|} \frac{1}{Rank_i} \qquad (10)$$

where *DGs* denotes the set of designated movies used in the evaluation, |*DGs*| represents the number of movies in *DGs*, and *Rank_i* is the ranking position of the *first* useful/relevant recommendation as determined by the appraisers.

As shown in Figure 11, *MR* outperformed Amazon and Redbox in terms of the MRR value, i.e., 0.65 versus 0.53 and 0.42, respectively, which are statistically significant based on the Wilcoxon Signed-Ranks Test ($p < 0.01$). These results verify that *MR* makes more relevant recommendations and ranks higher the relevant suggestions than the ones suggested by Amazon and Redbox, respectively.

5. Conclusions

Watching movies is one of the popular entertainments in the modern society, and these days people can watch movies anytime and everywhere—at work, at home, or in their cars. However, following the normal supply and demand curve, in the calendar year of 2016 up till mid-July, there were 7,547 most popular English-language movies released.[17] The increase in production of movies has created a problem for movie enthusiasts seeking new movies. In the year of 2018 the number of new movies released in the United States and Canada alone is close to a thousand [33]. Although websites with discussions on the latest and most popular movies are available, the amount of time needed to research movies has become insurmountable due to the large number of movies available. To decrease the amount of time needed to research personally appealing movies and help resolve the problem of needing to test movies out, we propose a novel movie recommender which suggests movie recommendations to its users based on a simple neural network model.

The neural network model is a computing system made up of a number of simple, highly interconnected processing elements, which process information by their dynamic state response to external inputs. A neural network is often referred to as an Artificial Neural Network (ANN), which has generated a lot of excitement in Machine Learning research and industry,

---

[17] www.imdb.com/search/title?count=100&languages=en&release date=2016,2016&title type=feature.

thanks to many breakthrough results in speech recognition, computer vision and text processing. We adopt the deep learning, which is essentially just deep artificial neural networks, to recommend appealing movies for moviegoers.

The proposed movie recommender performs well in terms of root mean squared error for collaborative filtering. When talking about collaborative filtering, we should clearly distinguish the following two tasks: (i) rating prediction and (ii) top-N recommendations. The task of *rating prediction* is much more popularized and, as a consequence, tons of papers and open source libraries are there. However, speaking about *top-N recommendation* task, the situation is quite the opposite, since in most business applications, it is required to compute top-N recommendations. Our work adds to existing literature which suggests that deep learning can be a powerful tool for a variety of problems in information retrieval [40]. In the end, this work makes improvement in terms of predicting ratings of and recommending top-N movies for users. Our recommender system applies regularization to further minimize the prediction errors. In addition, our system was able to handily outperform the neighborhood-based baseline, and was able to provide superior movie recommendations. Additional human assessments, which invoke college students and Mechanical Turk appraisers, further verified the relevance and usefulness of movies recommended by our deep learning model in terms of offering appealing movies for users to watch. As an added advantage of our deep learning approach, it is much more scalable at test time.

References

[1] Chai, T. and R. Draxler. 2014. Root Mean Square Error (RMSE) or Mean Absolute Error (MAE)? Geoscientific Model Development Discussions 7(1): 1525–1534.

[2] Changhau, I. 2017. LSTM and GRU—Formula Summary. https://isaacchanghau.github. io/post/lstm-gru-formula/, July 2017.

[3] Chervinskii. 2015. Autoencoder. https://en.wikipedia.org/wiki/Autoencoder, December 2015.

[4] Cho, K., B. van Merrienboer and D. Bahdanau. 2014. On the properties of neural machine translation: Encoder-decoder approaches. pp. 103–114. *In*: Proceedings of the Eighth Workshop on Syntax, Semantics and Structure in Statistical Translation (SSST-8).

[5] Chung, J., C. Gulcehre, K. Cho and Y. Bengio. 2014. Empirical evaluation of gated recurrent neural networks on sequence modeling. In NIPS Workshop on Deep Learning.

[6] Croft, W., D. Metzler and T. Strohman. 2010. Search Engines: Information Retrieval in Practice. Addison Wesley.

[7] Colah. 2015. Understanding LSTM Networks. http://colah.github.io/posts/2015-08-Understanding-LSTMs/, August 2015.

[8] Elkahky, A., Y. Song and X. He. 2015. A multi-view deep learning approach for cross domain user modeling in recommendation systems. pp. 278–288. *In*: Proceedings of the 24th International Conference on World Wide Web (WWW).

[9] Gal, Y. and Z. Ghahramani. 2016. Dropout as a bayesian approximation: Representing model uncertainty in deep learning. pp. 1050–1059. *In*: Proceedings of the International Conference on Machine Learning (ICML).

[10] Gers, F., J. Schmidhuber and F. Cummins. 1999. Learning to forget: Continual prediction with LSTM. pp. 850–855. *In*: Proceedings of the Ninth International Conference on Artificial Neural Networks (ICANN).

[11] Glorot, X. and Y. Bengio. 2010. Understanding the difficulty of training deep feed-forward neural networks. pp. 249–256. *In*: Proceedings of the Thirteenth International Conference on Artificial Intelligence and Statistics (AIS-TATS).

[12] Gomez-Uribe, C. and N. Hunt. 2016. The netflix recommender system: Algorithms, business value, and innovation. ACM Transactions on Management Information Systems (TMIS) 6(4): Article 13 pp. 1–19.

[13] Harper, F. and J. Konstan. 2016. The Movielens datasets: History and context. ACM Transactions on Interactive Intelligent Systems (TIIS) 5(4): Article 19.

[14] He, K., X. Zhang, S. Ren and J. Sun. 2016. Deep residual learning for image recognition. pp. 770–778. *In*: Proceedings of the IEEE Conference on Computer Vision and Pattern Recognition (CVPR).

[15] Im, I. and A. Har. 2007. Does a one-size recommendation system fit all? The effectiveness of collaborative filtering based recommendation systems across different domains and search modes. ACM Transactions on Information Systems (TOIS) 26(1): Article 4, pp. 1–30.

[16] Karatzoglou, A. and B. Hidasi. 2017. Deep learning for recommender systems. pp. 396–397. Proceedings of the Eleventh ACM Conference on Recommender Systems (RecSys).

[17] Koren. Y. 2009. The BellKor solution to the Netflix Grand Prize. www.netflix prize.com/assets/GrandPrize2009 BPC BellKor.pdf.

[18] Koren, Y., R. Bell and C. Volinsky. 2009. Matrix factorization techniques for recommender systems. Computer 42(8): 30–37.

[19] De Lathauwer, L., B. De Moor and J. Vandewalle. 2000. A multilinear singular value decomposition. SIAM Journal on Matrix Analysis and Applications 21(4): 1253–1278.

[20] LeCun, Y., Y. Bengio and G. Hinton. 2015. Deep learning. Nature 521: 436–444, May 2015.

[21] Liu, T., A. Moore and A Gray. 2006. New algorithms for efficient high-dimensional nonparametric classification. Journal of Machine Learning Research (JMLR) 7(6): 1135–1158.

[22] Liu, W., Z. Wang, X. Liu, N. Zeng, Y. Liu and F. Alsaadi. 2017. A survey of deep neural network architectures and their applications. Neurocomputing 234: 11–26.

[23] Pennington, J., R. Socher and C. Manning. 2014. GloVe: Global vectors for word representation. pp. 1532–1543. *In*: Proceedings of the Conference on Empirical Methods in Natural Language Processing (EMNLP).

[24] Qumsiyeh, R. and Y.-K. Ng. 2012. Predicting the ratings of multimedia items for making personalized recommendations. pp. 475–484. *In*: Proceedings of the 35th International Conference on Research and Development in Information Retrieval (ACM SIGIR).

[25] Rosebrock. A. 2016. A simple neural network with Python and Keras. Deep Learning, Machine Learning, pyimagesearch.com/2016/09/26/a-simple-neural-network-with-python-and-keras/, September 2016.

[26] Rumelhart, D., G. Hinton and R. Williams. 1986. Learning representations by backpropagating errors. Nature 323: 533–536.

[27] Salehinejad, H., J. Baarbe, S. Sankar, J. Barfett, E. Colak and S. Valaee. 2017. Recent advances in recurrent neural networks. arXiv preprint arXiv:1801.01078.

[28] Sarwar, B., G. Karypis, J. Konstan and J. Riedl. 2001. Item-based collaborative filtering recommendation algorithms. pp. 285–295. *In*: Proceedings of the 10th International Conference on World Wide Web (WWW).

[29] Shi, Y., M. Larson and A. Hanjalic. 2010. Mining mood-specific movie similarity with matrix factorization for context-aware recommendation. pp. 34–40. *In*: Proceedings of the Workshop on Context-Aware Movie Recommendation (CAMRa'10).

[30] Singh, A. and G. Gordon. 2008. Relational learning via collective matrix factorization. pp. 650–658. *In*: Proceedings of the 14th International Conference on Knowledge Discovery and Data Mining (ACM SIGKDD).

[31] Smucker, M., J. Allan and B. Carterette. 2009. Agreement among statistical significance tests for information retrieval evaluation at varying sample sizes. pp. 630–631. *In*: Proceedings of the 32nd International Conference on Research and Development in Information Retrieval (ACM SIGIR).

[32] Srivastava, N., G. Hinton, A. Krizhevsky, I. Sutskever and R. Salakhutdinov. 2014. Dropout: A simple way to prevent neural networks from overfitting. Journal of Machine Learning Research 15(1): 1929–1958.

[33] Statista. Number of Movies Released in the United States and Canada from 2000 to 2018. https://www.statista.com/statistics/187122/movie- releases-in-north-america-since-2001/.

[34] Tieleman, T. and G. Hinton. 2012. Lecture 6.5-rmsprop: Divide the gradient by a running average of its recent magnitude. In COURSERA: Neural Networks for Machine Learning 4.2: 26–31.

[35] Wang, H., N. Wang and D.-Y. Yeung. 2015. Collaborative deep learning for recommender systems. pp. 1235–1244. *In*: Proceedings of the 21th International Conference on Knowledge Discovery and Data Mining (ACM KDD).

[36] Wei, J., J. He, K. Chen, Y. Zhou and Z. Tang. 2017. Collaborative filtering and deep learning based recommendation system for cold start items. Expert Systems with Applications 69: 29–39.

[37] Wikipedia. Netflix Prize. en.wikipedia.org/wiki/Netflix Prize#cite note-netflixprize.com-21, May 2018.

[38] Yang, L., E. Bagdasaryan and H. Wen. 2018. Modularizing deep neural network-inspired recommendation algorithms. pp. 533–534. Proceedings of the 12th ACM Conference on Recommender Systems (RecSys).

[39] Yu, K., S. Zhu, J. Lafferty and Y. Gong. 2009. Fast nonparametric matrix factorization for large-scale collaborative filtering. pp. 211–218. *In*: Proceedings of the 32nd International Conference on Research and Development in Information Retrieval (ACM SIGIR).

[40] Zhang, S., L. Yao and A. Sun. 2017. Deep learning based recommender system: A survey and new perspectives. ACM Journal on Computing and Cultural Heritage (JOCCH) 1(1): Article 35, pp. 1–35.

Chapter **5**

A Recommendation System Enhanced by Topic Modeling for Knowledge Reuse in MOOCs Ecosystems

Rodrigo Campos,[1,]* *Rodrigo Pereira dos Santos*[2]
and *Jonice Oliveira*[1]

1. Introduction

The advancement of online education has revolutionized the way students learn around the world. The technological resources allow the analysis, optimization and availability of new learning options that benefit several users of such resources. One of these options is gaining more popularity: The Massive Open Online Course (MOOC), which emerged as a new educational philosophy. This advance is due to the reason it presents totally new definitions about the concepts of enrollment, participation, and even evaluation, but also for having values based on openness, ethics for participation, and collaboration[3].

With the highlight of MOOCs, the number of users has been growing constantly since its emergence in 2008. Several universities and other educational institutions have been adapting and reconsidering the classic learning structures and taking courses communities beyond the physical boundaries of the university with MOOCs [7]. From the interaction of several

[1] Federal University of Rio de Janeiro.
[2] Federal University of the State of Rio de Janeiro.
* Corresponding author: rodrigocampos.inf@hotmail.com

users around the MOOCs' platforms, some authors analyze the MOOCs from a perspective of software ecosystem (SECO), conceptualizing the MOOC learning community ecosystem [26], or simply MOOCs ecosystems. This perspective can (i) ensure more sustainable development for MOOCs, (ii) contribute with other benefits for the learning community, and (iii) bring partnership and alliance between universities, external companies, students and other ecosystems' stakeholders.

However, the growth of these platforms also creates some difficulties. The number of courses emerging in the ecosystem is increasing. Once learning institutes sometimes contain similar courses, there may be courses that address the same topics. This large number of courses can generate doubts when students should choose which course they will enroll. Therefore, some works consider the construction of courses recommendation systems for these students within a provider, as listed in the Section 2. In addition, there are some challenges for the recommendation in these scenarios and, if contemplated, they could facilitate students in accurately identifying content according to their learning needs, such as:

- A more personalized recommendation where there may be a merger of parts of courses. Currently, the recommendations consider the entire content of the course. For example, there is no assembly of a study plan, or some resource that might make the recommended item more flexible;
- Considering more than one provider in the recommendation;
- The recommendations do not usually merge the courses' data with other databases, either on the student or the items (courses) that are recommended;
- MOOCs have still interpreted the platform in a restricted way. Although there are several other actors interacting with this platform, these interactions are not mapped in the form of an entire ecosystem; or when they are, they still reflect the characteristics of a Virtual Learning Environment (VLE), which excludes a possibility of expansion and cooperation of the platform itself.

This work's contribution explores some of these major issues. It proposes an architecture of a web-based recommendation system that considers more than a single MOOC provider, enabling not only the full courses recommendation but also parts of courses in MOOCs ecosystems. The resulting system from such architecture aims to assist students in the process of searching for courses and to achieve demands and improvements, as well as sharing of software over the platform (i.e., reuse of knowledge).

In the Section 2, we describe how other researches propose related solutions and we indicate the main concepts of our work. In this chapter, some characteristics of SECO are addressed, exploring how these characteristics work specifically in MOOCs ecosystems. To do so, in the Section 3, a correlation between roles and a mapping of knowledge types shared in such MOOCs ecosystems is presented. Thus, it is possible to understand the importance of each stakeholder in the process of knowledge reuse. To build the recommendation system, two general steps are taken in this chapter: (a) identify the most used MOOCs providers, as well as which data is open for extraction, in the Section 4; (b) propose the web-based recommendation system, planning the different steps of the recommendation process and the knowledge reuse objectives for this architecture, in the Section 5. As the recommendation system includes topics modeling and labeling methods, the Section 6 address the most common concepts and techniques, as well as justification on the choice of techniques combination to be adopted in our system according to their relevance.

This chapter also includes an example of the recommendation system processing real-world data in the Section 7. This example allows to better visualize the whole process, from the user search in the recommendation to the issuing of courses and parts of courses recommended by the system. Finally, the Section 8 addresses the possibilities of the process extension and include some conclusions about the work and techniques.

2. Literature Review

Several studies have contributed to recommendation systems for MOOCs. They deal with different techniques to build and implement recommendations. The objective of the literature review in this chapter is to introduce the fundamental concepts of these techniques and approaches used in related work. The collection of material for literature review started by searching and analyzing recent work published in conferences about the recommendation system and/or education. Due to lack of material, we extended our literature review to cover Google Scholar indexed publications.

In order to classify the MOOC recommendation systems identified in the literature, the characteristics of a recommendation system raised in [22] were used as a reference. According to the authors, five aspects characterize the recommendation systems.

The first one is the Recommendation Technique, that addresses not only which recommendation technique is used but also how this technique is applied by the recommendation solution [22]. Recommendation techniques, to a greater extent, make use of two main entities for the recommendation process: user, which is the entity to which the recommendation is provided; and item, which is the product that is effectively recommended [1].

Among the studies identified in the review, Malakoudis and Symeonidis [20] suggest a recommendation system that applies Matrix Factorization (MF) as a recommendation technique. This technique has become practical in real-world scenario modeling because of its flexibility, since it can detect, from user ratings, how these users would rate items that have not yet been classified. From latent features, it is possible to sort items and make recommendations [17].

The Matrix Factorization technique is a class of CF approach of recommendation. This approach is applied in [29]—another work identified in this review that refers to the need to observe the history of the users to make the recommendation process in this approach. It is possible to find similar users' groups, represented by scores, and then predict the most appropriate items for a target user.

In contrast to the CF techniques mentioned before, the content-based methods consider descriptive attributes of the items, the so-called content (hence, the content-based name). These methods are used when there is no user's information, i.e., the ratings of other users are not known, as happens when applying CF. Thus, from a descriptor of item i, it is possible to find other items already evaluated with similar descriptors, given the level of similarity to recommend or not recommend the item i [1]. As such, Case-Based Reasoning (CBR) has played an important role in content-based systems. A leading work that uses the CBR technique is presented in [8], based on a principle that "similar problems have similar solutions", and treat problems and solutions as cases stored in a library called as Case Base.

The second aspect is the type of recommended items. It refers to the characteristics that involve the type of content that the recommendation system recommends to users [22]. When it comes to recommending courses for users, MOOCRec [8] and MOOCRec.com [20] use recommendation approach for this purpose.

The latter aims to help students in the acquisition of skills that are expected from their ideal job through a successful recommendation.

Differently, OERecommender project [29] recommends Open Educational Resources (OER). Similarly to MOOCs, OER is a concept that is part of Online Open Education, one of the most important movements for education in the 21st century [3]. Even though they are part of the same movement, they involve different concepts. OERs consist of 'any kind of educational material in the public domain or associated with an open license' [3]. Meanwhile, MOOCs are defined as 'online courses accessible to anyone on the web' where 'institutions have joined in an effort to make education more accessible by teaming up with MOOCs providers' [12].

The third aspect is the Output Form, i.e., information about how the user receives the recommended contents, such as system-driven or automatically provided by the system as a facilitator for the user [22]. In OERecommender project [29], the recommendation is automatically triggered by the system, by a search engine that captures the metadata and generates CAM instances from a MOOC course that the student is learning via the browser.

Meanwhile, MOOCRec.com makes recommendations based on user search. It makes use of crawlers to capture items that are recommended to users according to the information given in the search at runtime. The same happens with MOOC-Rec that considers the interests of the user through an interface where it can do a search for suggestions. The system translates this query into a query and the output is returned in this same interface.

The fourth aspect described is the Cross-Dimensional Features, that contain characteristics of the system components—if any component of the recommendation system applies any specific technique, such as considering user feedback [22]. In OERecommender project [29], a future work proposals is to extend the recommended through a prediction based on machine learning. The machine learning approach has been used in some works of recommendation systems, as in [2], which combines data mining algorithms, such as clustering and association rule to recommend courses in Moodle e-learning. Although it is not a solution for MOOC, the work emphasizes that this combination can be applied in MOOCs. The results show that the Simple K-means clustering and Apriori association rule algorithm would be the most suitable for this recommendation scenario since it is not necessary to have a data preparation stage and the number of association rules is bigger.

Another proposal that may be considered as related to our study is MOOCLink [12]. This system is not a recommendation system, but rather an aggregator, since it integrates different MOOCs courses' providers, adding courses to facilitate their search and comparison. To make the clustering possible, it uses LOD. Although MOOCLink does not apply recommendation, LOD has already been used to support the recommendation in other works with good results. For example, Di Noia et al. [13] suggest a content-based recommendation system with linked dataset exploration of open data in the scenario of movies, such as DBpedia. The work also contributes to the identification of similarity in these bases and it allows item recommendations with the trained system.

In the CF class recommendation system, we have identified the Heitmann and Hayes [15] proposal that uses LOD to increase what the system knows about new users or new items. It helps to solve the cold-start problem, which is very common in open recommendation systems, happening when the system has little information about a new user, a new item, or when it did not have many interactions.

Some components proposed in MOOCLink [12] and in [15] were adapted for application in our proposal, such as the application of an integration service using the Karma Web to integrate data from several sources into the RDF (characteristic of LOD), as described in detail in the Section 5.

The last aspect identified is Architecture. It refers to the implementation status of the system, which involves details such as whether the proposal is already available to be used (or not) until part of the process has been developed. It also includes information on allocation and access, i.e., how the contribution is materialized (e.g., web-based tool, web-based architecture, desktop application) [22]. Among the analyzed works, only the OERecommender project [29] differs in this aspect, being a web widget that allows the collection on information of the user when added to the browser, making searches and showing the recommendations of OER according to the algorithm. The other works—MOOCLink [12], MOOC-Rec [8], and MOOCRec.com [20]—present solutions based on the web, and a user can access them through a URLs.

Based on information from related work and according to the characteristics of the recommendation systems. Table 1 summarizes the most related systems, namely: MOOCLink [12], MOOCRec.Com [20], MOOC-Rec [8], and OERecommender [29]. Meanwhile, our proposed Web-Based Recommendation System is referenced with the acronym "WBRS" in the last column.

Regarding the status of each project, [29] and [8] have not implemented their solution yet. [20] have implemented their solution and it is available for use. MOOCLink [12] is finished but it is not deployed on any web server for use (i.e., it is not running).

As observed in Table 1, our work differs from others in some features. The effective contribution of our work is the creation of a recommendation system applied in the context of MOOCs. In this context, it is possible to include scientific contributions involving the recommendation process and the development of the work, such as part of the courses' recommendation in addition to whole courses, delivering the users' packages of courses according to their knowledge gap.

One of the contributions regarding the recommendation technique is the use of an approach called "hybrid recommendation", which is a combination of CF and content-based filtering. This approach makes use of a Machine Learning algorithm, more specifically Topic Modeling, which group course topics to identify similarities between them and optimize the recommendation process, as presented in the Section 6.

In addition, the application of LOD to the collected background data in our proposal allows the construction of a recommendation with more advantages regarding the use of crawler or CAM (Contextualized Attention Metadata). Another important differential of this work is the definition and analysis of MOOCs within a larger context, called MOOCs ecosystems

Table 1: Comparison of related work and the web-based recommendation system (WBRS) architecture proposed in our research (RS = the concept of recommendation systems).

		[12]	[20]	[29]	[8]	WBRS
Recommendation Technique	RS related to CBR				X	
	RS designed with CF			X		
	RS uses MF		X			
	Hybrid					X
Cross-Dimensional Features	Linked Data	X				X
	Machine Learning					X
	Slop One			X		
Type of Recommended Items	Courses		X		X	
	OER			X		
	Courses and its parts					X
Output Form	System-driven		X		X	X
	Automatically provided			X		
Architecture Status	Not Implemented			X	X	X
	Finished but not running	X				
	Running		X			
Architecture Interface	Web	X	X		X	X
	Web Widget			X		

conceptualized from the SECO concept that contributes to the information reuse and integration based on actors' interactions in the ecosystems. We also consider the fact that all the related projects have a multi-provider approach, but only our proposal presents a recommendation for parts of courses (as well as complete modules or courses).

3. MOOCs Ecosystems

The SECO perspective for the MOOCs platforms has still been little explored. The discussions pointed out in [26] or in [9] highlight the difficulties in identifying the specific functionalities of MOOCs as a barrier to understanding the MOOCs learning community ecosystem. It happens because the research on MOOC still deals with the general characteristics of virtual learning community ecosystems. Thus, the actors' roles and their relationships need to be defined at different stages of provider utilization. The SECO concept also considers that these relationships are supported by the technological platform of the MOOCs (or technological market of MOOCs). This business model functions as a unit and operates through the exchange of information, resources, and artifacts [16].

This broader vision of MOOCs providers brings several benefits to those involved in addition to sustainable development. We can be mention that MOOCs: (a) facilitate innovation, knowledge sharing, and software evolution; (b) strengthen cooperation in its multiple and independent entities; (c) increase the attractiveness of the platform, bringing new players to the ecosystem; and (d) assists in choosing the best platform, through the identification and analysis of software architecture, mapping product design, business tasks, and risks [4].

To define these actors' roles and their relationships, it is necessary to understand the activities and responsibilities of each role in the SECO concept. To do so, different concepts about the actors' roles in ecosystems are investigated in [18]. These categorizations are adapted and presented in Table 2.

Table 2: Description of SECO actors' roles. Source: (Lima et al.2016).

Hub	Keystone	Adds value to SECO and is primarily responsible for maintaining health, i.e., longevity and growth. It can represent the dominant entity of influence.
	Dominator	Extracts value from SECO, putting its health and sustainability at risk.
Niche Player	Customer	Represents the customer, who generated the need for the SECO software products.
	Competitor	It tries to extract value from the ecosystem but does not threaten the SECO's health.
	Supplier	Actor providing one or more products or services required by the ecosystem.
	Vendor	Sells SECO software products. Can be classified as Reseller, Independent Software Vendor (ISV), or Value-added Reseller (VAR).
	Developer	Internal developer linked to SECO formative entities, being classified as Influencer, Hedger or Disciple.
External Actor	3rd-party developers	Promotes SECO and its products, and can propose improvements; similar to Influencer, but external to SECO, having no formal bond with Keystone.
	End-user	Product's final user, but differs from Customer for not hiring Keystone service.
	External Partner	Contributes to the SECO well-being through attitudes, such as the promotion of SECO and its products, also proposing improvements.

For the categorization of roles in MOOCs ecosystems, it is necessary to observe the platforms' basic structure and their actors. Since MOOC is a novel technology, there is still no consensus regarding the groups of actors involved. In [5], the most prominent groups of actors are teachers, students, private actors (e.g., advertisers or employers), and higher education institutions. In [11], the authors describe two more actors (course designers and manager), refer to teachers as tutors, and refer to students as learners, i.e., they do not consider only the actors with more participation in the platforms. In [26], there is also another change from that one addressed in [11]: the course designers actor is mentioned of "those who make MOOC", i.e., in a more generic way.

For the conceptualization of the MOOCs ecosystems, it is possible to correlate these groups with the predefined SECO roles, considering that each group on the platforms performs some functions and interacting with certain groups, where both can exchange information. This correlation is presented in Table 3, which also contemplates three different stages [26], where each role can be played by different roles in each of the stages:

- In the first stage, students use an email and some personal information to register with the MOOC provider. At that point, they create a new account that can be used to log in for the first time and finally sign up for new courses. Although they can acquire a new product from the platform (courses), they still lack enough knowledge for any kind of interaction with other MOOC users. This makes these students play the role of consumers [26];

- In the second stage, student interactions happen in a separate way, i.e., part of it within the platform, in the existing forums and discussions; and another part outside the platform, seeking knowledge from other sources, downloading materials from the internet, editing and producing the own material based on internet content, and sharing this kind of knowledge with other users in the learning network. This knowledge can be any personal resource, process or personal learning notes, and they are shared in forums, wikis, email or any other means of interaction. As such channels disseminate knowledge of MOOCs ecosystems based on inserting an external knowledge based on personal perceptions in the network, students can be considered as decomposers [26], which in a perspective of SECO is equivalent to a dominator, since these are responsible for extracting the maximum value from the ecosystem, destroying it [28];

- In the third and last stage, the knowledge absorbed by the students allows them to assist new students in the learning process, collaborating with the community as they are already able to deal with doubts and learning difficulties, as well as usability issues. With these characteristics, in this

Table 3: Relations between SECO's roles and MOOCs ecosystems' roles.

	1st Stage	2nd Stage	3rd Stage
Keystone	Higher Education Institutions		
Dominator	–	Students	Students
Customer	Students		
Competitor	Advertisers		
Supplier	Teachers, Course Designers	Teachers, Course Designers	Teachers, Course Designers, Students
Vendor	MOOCs Providers (ISV)		
Developer	Course Designers		
3rd-party Developers	–	–	–
End-user	–	–	–
External Partner	Employers		

stage, students increase the community's strength and can be considered as suppliers [26].

Knowledge exchange based on existing interactions can be mapped from the identified roles. In this context, it is possible to emphasize the difference among the concepts of data, information, and knowledge. Data are simple facts that become information, if organized into an understandable structure. Meanwhile, the information can become knowledge when, from a cognitive processing/validation, can fit in a context as a result of this process, besides being able to make predictions [10].

In MOOCs ecosystems, the knowledge exchange happens based on the connections between the mapped roles, where information is exchanged in the means provided by the MOOC provider itself, such as forums, chats, wiki, and others. Even logs are a form of information exchange in these ecosystems [26]. Table 4 consolidates technical information from the providers themselves and some works in the literature that explore the main interactions between actors in MOOCs ecosystems, allowing to better visualize the importance of each connection among providers.

Each interaction presented in the first column of Table 4 can be represented in the graph in Figure 1, where a node represents an actor and edges are the interactions between them. The only node that is connected to all other actors is the "MOOC Provider" identified in the lower right corner of Figure 1. Another detail that can be observed in the graph is the direction of the edges that is equivalent to the direction of the arrows of the interactions identified in Table 4.

As the course is absorbed by providers (who store, process and show this information), there is a dependency between the student and the provider,

Table 4: Interaction between different groups of stakeholders.

Interaction	This Interaction Exists. . .
Students → MOOCs Providers	to help students follow courses taught by teachers.
Students → Higher Education Institutions	to help students improve their employability, looking for information on the course quality.
Students ↔ Employers	because students may exercise their abilities with employers from the ecosystem who, in turn, have access, via a MOOC platform, to a large pool of students as well as to detailed data about their skills.
Students → Advertisers	if the advertiser's presence and their payments allow platforms to offer courses to students for free.
Students → Students	because students might be influenced; as a result, student learning outcomes depend on interactions with fellow students.
Teachers → MOOCs Providers	because teachers seek to disseminate their teaching materials and experiment with new pedagogies.
Teachers → Higher Education Institutions	because even if teachers can offer a course in their own name, they usually still depend on their respective university.
Teachers → Employers	because teachers value employers' presence indirectly if they contribute to attracting more students.
Teachers ↔ Students	because they can interact with each other via the MOOCs' platform, by social media, or by telephone, meeting and answering activities in real life. Currently, students have organized offline meetings.
Higher Education Institutions → MOOCs Providers	because institutions can decide to invest money and time in a MOOC platform.
Higher Education Institutions → Teachers	since institutions pay teachers and encourage them via other non-monetary rewards.
Higher Education Institutions → Employers	because institutions only value the participation of private actors to the platform indirectly.
Employers → MOOCs Providers	because employers see MOOCs as a flexible and cheap tool to train their staff.
Advertisers → MOOCs Providers	since advertisers are ready to pay before having access to the platforms' visitors, as well as information about them.
Course Designers → MOOCs Providers	because courses are designed and published in the MOOCs providers' platforms.

since whenever he/she interacts with the platform, the MOOC provider will make such information available. Moreover, given the large amount of data that is exchanged, the Student → MOOCs Providers interaction is the one that most generates knowledge for the ecosystem. This extracted knowledge comes to the student in text format, video, games, audio, animations, blog, chat, forum, e-mail, or even virtual communities.

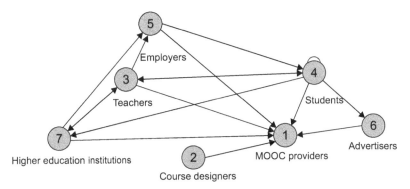

Fig. 1: Interactions in the network of actors within the ecosystem.

However, in addition to the extraction of knowledge from a provider, a student can also provide knowledge. Some actions, such as assisting other users in the forums, submitting a response to a proposed activity, attending classes in each course or even signing up to the system can generate knowledge for other actors. In this context, providers are responsible for storing such information in databases and making it available to other actors, such as higher education institutions or teachers in the processes of student assessment and querying (explicit knowledge), or for management, statistical data extraction and decision making about courses' pedagogical plans.

4. MOOCs Data Extraction

In order to define a recommendation architecture to support multiple providers, it is necessary to map the level of data openness, as well as the possibilities for obtaining such data. To do so, this section maps data from the most common providers based on the technical literature, as shown in Table 5.

In order to choose which provider could be used in the referral system based on the data openness, it was identified that only the edX provider API does not have a totally open availability of the data to use, since it requires an OAuth authentication, as described in [12]. Meanwhile, the other providers (Coursera, Udacity, Khan Academy, and OCW) hold free access and were selected to the recommendation system.

Other information that can be considered in Table 5 is the type of information that can be extracted from each server and the uniqueness of data format. All APIs allow the extraction of data in JSON format, which makes it easier to integrate with the recommendation system, whose architecture is proposed in the next section.

Table 5: Information about provider extractions.

	How to Obtain Data?	It is Possible to Extract. . .	Data Format
Coursera	Coursera API	all of Coursera's courses, instructors, and partnering universities	JSON
edX	Crawler	limited information	Several
	edX API	Courses API, Data Analytics API, Discussion API, Enrollment API, Grades API, User API, Discovery API	JSON
	RSS Feed	a list of edX course list	XML
Udacity	Crawler	limited information	Several
	Udacity API	course catalog information and nanodegree courses	JSON
Khan Academy	Khan API	"topic tree" which gives the entire hierarchy of Khan Academy's course offerings. It can also obtain the list of all badges, badge categories, details of a particular course, etc.	JSON
OCW	OCW API	indexes of all these courses (e.g., links, hash, provider, language, tags, author, title, description, published, indexed, modified, categories)	JSON
	Excel Dump	all the courses (e.g., links, hash, provider, language, tags, author, title, description, published, indexed, modified, categories)	Excel

5. Proposed Architecture

In this section, we address a recommendation system architecture and its stages, from the providers presented in the Section 4. To build a collaborative and open recommendation system, we opted to make use of Linked Data from the data integration approach of several MOOCs providers exposed in [12]. The use of this integration technique is optimized in our work. Another approach is the modeling and labeling of topics, detailed in the 6, which explore techniques that can improve data representation and facilitate the identification of courses' central topics. The architecture model with these elements is presented in Figure 2.

The most benefited actor in the recommendation process presented in this chapter is the student, since courses and part of courses are recommended to them. The impact of a well-made recommendation is directly connected to better satisfy these students with the recommended content, which may influence them not to leave the courses. Consequently, it has the effect of

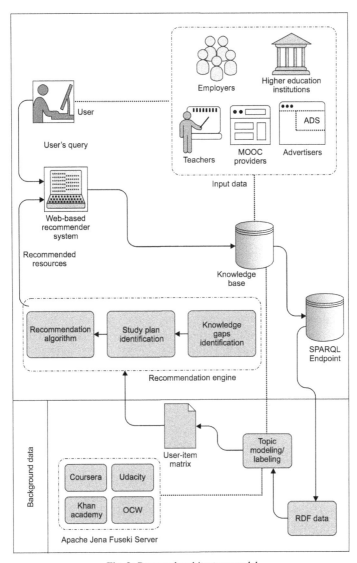

Fig. 2: Proposed architecture model.

reducing dropout rates in providers and the student's existing knowledge gap. In general, these benefits positively affect other roles.

It should also be observed that the approach of an open recommendation system (i.e., open data supporting the recommendation process) was adopted precisely to consider the multiple data sources of the MOOCs providers chosen. One of the benefits of using a variety of sources is to alleviate common problems in recommendation processes, such as cold start (i.e., the problem of recommending a course when no data is available about the item or user) in approaches that use Collaborative Filtering (CF). The enrichment of recommendations and the reduction of low user rating problem [15] can also be cited as benefits.

The flow of the recommendation process is determined by the direction of the arrows in Figure 2. Dashed lines represent actions that are performed outside the main stream. The entire process can be organized into some layers and steps:

- A user submits a refined and textual search through a series of search options in the Web-Based Recommender System. To make the recommendation, the system searches the Knowledge Base, selecting specific data from a student (e.g., curricular history), general data of the actors related to that student in the ecosystem (e.g., information from the universities where the student took the courses), and other information inferred by the recommendation system (e.g., user's competencies). This information is part of the Input Data, i.e., information that helps in the process of recommending a given user and which represents the knowledge acquired from interactions in the ecosystem. As the student is the main beneficiary of the recommendation process, the step illustrated in Input Data box in Figure 2, contemplates the actors who have interactions with students (i.e., just course designer is not included);

- Next, the SPARQL Endpoint is initiated. This semantic web technology allows the Resource Description Framework (RDF) Data model to be searched in different schemas;

- Heitmann and Hayes [15] introduce some details on how to integrate Linked Open Database (LOD) into recommenders. Although they have applied in a different scenario, the components used can be pointed to integrate the data in our Knowledge Base. Two components are defined: the data interface for capturing RDF data, using the HTTP protocol to access the URIs; and the integration service to match the representation of data from different sources. These procedures are executed in our architecture by an integration tool called Karma Web [27], which allows these procedures to be organized easily and quickly. The integration process works by informing which MOOC API is used to the Karma Web. This process is possible since the data from the chosen MOOCs providers are in the JSON hierarchical format. Quickly and intuitively, data is converted from raw JSON to RDF data. The choice of providers, choice motivation and analysis of what can be extracted are presented in the Section 4;

- With data in RDF, the next step in the architecture process is the topic modeling method that integrates that RDF data representing the knowledge base with the background data and transform it into a user-item matrix, where the user column contains information from the knowledge base and the background data represents the item. These procedures are detailed in the Section 6. This step also introduces techniques for labeling topics and arranging these topics in the user-item matrix;

- Finally, the user-item matrix is read by the recommendation engine procedures, which also is supported by the knowledge base data for process enrichment. First, Knowledge Gaps Identification collects the user's skills as well as desired qualities by defining what the system calls a knowledge gap. The system then uses the matrix and course information to identify which courses or parts of courses may be enough to fill the student gap. The stage responsible for this is the Study Plan Identification. Then, the last step is to use these courses and identified parts, ranking them according to a criterion of relevance in Recommendation Algorithm, so just send them back to the user.

In addition to the benefits to students in their choice processes, the proposed architecture presents features capable of reaching the four stages of knowledge reuse defined in [21]: (a) capturing or documenting knowledge; (b) packaging knowledge for reuse; (c) distributing or disseminating knowledge (providing people with access to it; and (d) reusing knowledge.

As shown in Table 4, MOOCs ecosystems focus on a shared management in order to strengthen stakeholder collaborations, considering that software asset management is decentralized in a SECO. We can cite as an example the fact that in Coursera any user of the community can create a course and this course can be made available on the platform at the same time. Higher education institutions also register development partnerships with suppliers. This means that the ecosystem perspective allows stakeholders to participate in the management of this reusable knowledge by consuming, providing or in any other activity explained in Table 3.

These tools consolidate inputs that benefit diverse ecosystem actors with reusable knowledge, as well as grouped, stored, and user feedback. Moreover, an advantage could be automatically identifying new course demands, consequently affecting contributions to partnerships and alliances, if a common interest demand is identified. Further advantages include enhancements to the existing content or even software improvements indicated by providers.

6. Modeling and Labeling Topics

To fill in the user-item matrix, the technique of topic modeling is used to identify which are the central themes linked to the user and to the item of this matrix, as well as to facilitate the representation of these themes for use in the recommendation engine. The most common topic modeling processes currently work as follows: from a collection of documents $D = \{d_1, d_2, \ldots, d_{|c|}\}$ and their respective fixed vocabulary of words $V = \{w_1, w_2, \ldots, w_{|v|}\}$, they distribute all of these words (represented by terms w) into groups (represented by topics θ) with their probability $p\,(w|\,\theta\,)$, which gets higher each time that term is more related to a topic. Then, through the probability distribution $p\,(\theta\,|d)$ technique, it is verified the probability of each document being linked to each topic. Thus, an association between topics and documents is created [25].

However, these topics are not always easy to understand. This can cause a loss in the part of the topic modeling goal regarding the identification of documents' central theme. This can be due to a lack of understanding of the enumerated terms, either because the lack of domain knowledge or the difficulties in choosing a single theme among many words. To address such problem, there are topic labeling methods that seek to select a word (called label) to express the theme or topic area [19].

Nolasco and Oliveira [24] describe some known techniques for applying topic modeling and labeling in practice. In addition, it is analyzed the possibilities of its application and where it is most used. Tables 6, 7 and 8 give a brief summary of the techniques addressed in [24] and their respective definitions.

Although LDA is still one of the most used techniques for topic modeling, the state-of-the-art presents methods that use LDA as a base but modify basic assumptions to better represent data according to the application domain and its particularities. In order to insert a practical and better-organized representation for the recommendation system proposed in this chapter, we seek to select the best techniques of topic modeling from those presented in Table 6. Moreover, the most appropriate techniques of topic labeling are selected from the presented in Tables 7 and 8.

We must consider techniques that best suit the fact that the recommendation system proposed here uses a massive amount of data and that this mass of data is constantly modified, since courses are often created and/or excluded in the MOOCs providers. The representation of topics and labels across different domains of MOOCs courses would become less practical in a manual way. To meet these needs, specific techniques have been chosen which are presented in Figure 3. The technique chosen in block 2 (Topic Labelling) is justified by the fact that text-based approaches present better results with such technique than with other existing ones [23].

In addition, the combined use of these techniques addresses the needs discussed above, ensuring a better labeling process. Block 1 (Topic Modelling) in Figure 3 also presents an optimized technique. The LDA [6], which

Table 6: The most well-known techniques for topic modeling.

1. Topic Modeling	Description
Latent Semantic Analysis (LSA)	It uses linear algebra with SVD (Singular value decomposition) to decompose a corpus into its subjects. LSA is used to categorize documents, search for documents by keywords, and generalize results through similar documents in other languages.
Latent Dirichlet Allocation (LDA)	Distribution of groups for each term of a textual document and a distribution of groups for each document. Thus, one can group the documents according to the probabilities associated with each group.
Extended Topic Models	They apply the LDA method in order to expand the basic assumptions of the method, increasing the possibilities of application and improving results.
1.1 Implementation libraries	**Description**
lda-c	C Language
Mallet	Java Language
Gensim	Python Language

traditionally requires the number of topics to be extracted as a parameter to be inserted, is combined with the stability analysis approach described in [14], which makes it possible to automatically infer the topics value for a collection of documents.

Once the techniques have been chosen, it is necessary to define how each modeling and labeling stage is performed in the process of generating the user-item matrix. As the proposal of this system is also to recommend part of courses in MOOCs, in addition to whole courses in these providers, the matrix item is a vector with those modules and parts of the courses of all the collected providers (as presented in the Background Data in Figure 2). This vector is organized and divided into topics. Each topic represents a theme or learning area. Therefore, for modeling these topics, each course module is a document with specific content.

The "Topic Modeling" (box 1.1 in Figure 3) step is responsible for applying the LDA algorithm to process the terms of these document and, through the distribution of groups, performs the separation of these terms according to the themes. The default LDA method requires the desired number of topics to be entered as a parameter. Since the number of themes in MOOCs environments is large and considering that new themes may emerge as new courses emerge, the automatic method to number of topics [14] is integrated in conjunction with the LDA to automatically infer the number of topics based on the stability of words on top of the multinomial distribution of each topic. The outputs of this modeling process are the topics with their

Table 7: The most well-known techniques for topic labeling.

2. Topic Labeling		Description
Semi-supervised Approach	The combined use of Ranking	Automatically generated labels based on expert collection training.
	Active Learning	The system extracts terms to represent the area in a simple way and the experts give feedback until the system sets and the term is satisfactory to be labeled.
Automatic Approach	Own list	Apply a simple criterion in the list of terms (e.g., the 10 most relevant).
	Statistic over all the words	The system applies some statistics from all the words in the collection.
	Combination of List + Statistical Process	They apply a statistic but considers the list of terms already organized in topics.
2.1 Selection of Candidates		
A sample of the most relevant documents		Through the associated probability, the system discovers which are the most relevant documents, not having to use all the documents for the process.
All content in the collection		The whole set of documents is considered.
Content + external databases		In addition to the set of documents, they can use external databases, such as Wikipedia and Ontology.
2.1.1 Text Extraction		
Textual		Terms are extracted from the text body.
Keywords		Through classifications, it aims to extract terms defined by the author to describe the whole set of documents.
Natural Language Processing (NLP) extraction		Extraction of nominal phrases using PLN.
Keyword extractor		Build an extractor to select keywords.
Based on fast keyword extraction algorithm		Select all the words that are between stop words and phrase delimiters (such as a comma).

respective lists of terms with a greater probability related to the topic. As LDA uses all the vocabulary for the probability distribution, each list contains all the vocabulary with the respective probabilities that vary according to the proximity of the relationship between term and topic. The Implementation Library chosen in this system was Gensim, from the Python language (box 1.1 in Figure 3).

By showing the terms best placed in the probability list of each topic, it is difficult and time-consuming to define a topic label manually. Therefore, the Topic Labeling technique uses the topic list itself as input to the process. At the same time, it uses a statistical method that saves time in the labeling process and allows data scalability. This method, which corresponds to the "Selection of Candidates" (box 2.1 in Figure 3), is characterized by using only a sample

Table 8: The most well-known techniques for topic labeling (ranking and label selection).

2.2 Ranking	
Term Frequency (tf)	Assigns points to terms based on relevance, i.e., how often these terms appear.
Term Frequency-Inverse Document Frequency (tf-idf)	In addition to the frequency of tf, they apply a method that can prevent stop words from receiving high scores.
Degree/Term Frequency (deg/tf)	Consider how many times the term appears isolated but also considers how many times it appears in a term.
Modified Label Degree (mdeg)	It considers the number of occurrences of a candidate label in the list of labels but also considers how often that label appears in other candidates by assigning different weights in the formula.
2.3 Label Selection	
Individual Selection	Select only the first label in the list.
Inter-topic Selection	Applied when the same label appears in two topics.
Intra-topic Selection	Applied to select the best sequence of labels within the same topic, aiming to facilitate the understanding.

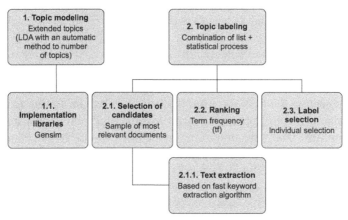

Fig. 3: Techniques of topic modeling and labeling chosen.

of more significant documents in the collection, not considering all the terms in the lists. This is possible because there is an associated probability (i.e., the probability of the words in the text of a document being associated to each topic) in the relationship between Topics and Documents. Thus, the stronger the association of a document to a topic, the more relevant that document is to that topic. To select the list of candidate labels for a topic, the technique chosen in this chapter uses only the relevant TOP D documents of that topic according to the associated probability. D indicates the number of documents, which is freely estimated in the application.

With the documents properly selected, the next step is to perform the "Text Extraction" (box 2.1.1 in Figure 3), which consists of implementing the Fast Keyword Extraction algorithm. This algorithm has as input a list of stop words and phrase delimiters (such as commas). By having the documents iterated, the algorithm checks all the words that are between such stop words and delimiters, considering this text as primitive labels, i.e., they are not candidates yet. This is because there is still a second check that is not in fast keyword extraction. To select the most relevant words of TOP D documents, the text extraction steps check if these primitive labels are contained in the TOP W terms of topic θ. If they are, enter the list; otherwise, check the next ones until it finds one that is contained. The variable W corresponds to the number of terms and, similarly to D, this value is estimated freely in the application.

The next step is the "Ranking" (box 2.2 in Figure 3), where the technique TF (Term Frequency) is applied. The objective is to assign points to each candidate according to the relevance that this term has. In this case, the relevance is defined by that term's frequency of occurrences. Since the stop words have already been deleted in the previous step, there is no need to apply IDF, i.e., a technique that is usually worked together with TF to exclude stop words.

The last step is the "Selection of Labels" (box 2.3 in Figure 3). As the candidates were ranked in the previous step, the options already appear in order of relevance, which means that we have a term that represents the topic well by selecting the first one from the list; so, it can be considered a label. This is also possible because we want to select an individual label. In the case of multiple label selection, some adaptations would have to be made and another technique should be adopted.

Therefore, the results of this last process are the topics θ (each of them with its respective list of terms R) and a label l for each topic. These topics represent the item in the user-item matrix, as shown in Table 9. For better reading by the recommendation system algorithm, this list is transformed into a vector notation. Then, we have a vector of discipline syllabus (θ_i) and their respective areas (l_i). To identify the user of the matrix, we must construct another vector. This vector represents what each user knows about each of the identified areas. To do so, we need a new collection of documents $D' = \{x_1, x_2, \ldots, x_{|y|}\}$, where each document x_i is represented by a "curriculum", i.e., a document from the Lattes[1] curriculum, a document from the LinkedIn[2] curriculum, or even the student completed courses information in the selected

---

[1] An integrated system maintained by the Brazilian government to manage information about researchers. Available in: http://lattes.cnpq.br/.
[2] Available in: https://www.linkedin.com/.

Table 9: User-item matrix notation.

	Item	User
Topic–Label	Terms (Discipline Syllabus Vector)	Vector of Documents (Curriculums)– Associated Probability
$\theta_i - l_i$	R_i	$x_1 - p\,(\theta_i|x_1),\, ...,\, x_n - p\,(\theta_i|x_n)$

providers. This information is contained in the Knowledge Base system (as shown in Figure 2). Thus, the LDA method is applied again, but this time only with a distribution of groups for each document in that collection.

Considering the existing topics in the matrix, it is not necessary to apply the method for inference in the number of the topics again. The distribution of groups then generates the relationship between documents (curriculum) x and topics (of elements) θ with their respective associated probabilities $p\,(\theta_i|x_j)$. The more associated the words of a document with the terms of a topic, the greater the associated probability, and consequently the more relevant that document is for the topic in question. In this context, the system distributes all documents in the user of the topic area with the associated probability. This distribution is also be given in vector format. Since the number of curriculums is much smaller than the number of discipline syllabus, several areas have an associated curriculum in the matrix with the associated probability equal zero. The notation for the complete matrix is shown in Table 9.

With the item vectors, user vectors, and label for distribution, the recommendation system has more organized data and ranked elements to perform the Knowledge Gap Identification, Study Plan Identification, and then the recommendation of the best course module options to students. The prediction techniques of the recommendation system allowed, for example, filling in the user column in Table 9 that has zero or low associated probability, i.e., where the student does not have curricular experience with inferences. The system understands that the higher the degree of inference, the greater the student's interest in this content, and it is possible to recommend the elements of that area in an organized way.

7. Example of Use

This section explores an example of how the system would perform in a situation where a user wants to know the best course/module to be done, given a preview interest in a knowledge area. This example of use helps to understand the idea of our solution and provide a preliminary evaluation of how the recommendation system works.

In the present example, the student with a dummy name (Tom) accesses the recommendation system after having completed the MOOC course "Intro to

HTML/CSS: Making web pages"[3] from the Khan Academy provider. During the course, Tom realized that the classes contained some Structured Query Language (SQL) commands, but that the course did not address any further explanation on the subject. When doing some searches with the word "SQL web" in this provider, Tom visualized that there are several courses addressing the subject, but that in his perception none would be appropriate according to his interest. Therefore, Tom's intention in accessing the recommendation system is to receive recommendations on what would be the best course/module that could fill this knowledge gap.

Upon accessing the system, Tom searched for the words "SQL web", just as he had watched on the Khan Academy platform. The system allows Tom to select some search filters, that were: Start Soon (for availability), Introductory (in level), English (as the language), and Free (for value). The moment Tom submits his search, the first process of the recommendation system is to retrieve the input data from the user to the knowledge base. The key information retrieved at this stage is the student curriculum information, such as course history from other providers, Lattes curriculum, and LinkedIn curriculum. Then, SPARQL groups all the course information and its menus contained in the Background Data layer. Since this layer contains information from four different providers, including the Khan Academy, Tom's previous search results are also grouped together for data integration. Through the Apache Jena Fuseki Server, the raw data is transformed into an acceptable RDF, creating and maintaining the SPARQL endpoint and then executing SPARQL queries according to the search submitted by Tom.

To populate the user-item matrix, the Topic Modeling step organizes the search results of SPARQL in the item column, as shown in Table 10. Then the Topic Labeling step generates a single label for each topic, filling the first column of the matrix, as shown in Table 10. Then, information previously retrieved by the Knowledge Base is processed only by the Topic Modeling step, which groups Tom's curriculums according to the course topics grouped in the item. This step populates the user column of the matrix, as shown in the last column of Table 10. With the matrix filled in, the information can be sent to the recommendation engine, which in turn makes predictions based on user columns where the user has low or zero associated probabilities, being able to predict which course or parts of courses would match the student's needs.

From this functionality not just recommending complete courses, it was possible to recommend the book "SQL for Web Nerds"[4] which is part

---

[3] https://pt.khanacademy.org/computing/computer-programming/html-css.
[4] https://ocw.mit.edu/courses/electrical-engineering-and-computer-science/6-171-software-engineering-for-web-applications-fall-2003/readings/.

Table 10: Examples for illustrating attacks.

	Item	User
Topic–Label	**Terms (Discipline Syllabus Vector)**	**Vector of Documents (Curriculums)–Associated Probability**
θ_1–Database	{SQL for Web Nerds, Introduction to Databases and Basic SQL, Advanced SQL, Accessing Databases using Python}	Lattes –0.89, Curricular History –0.75, LinkedIn –0.71
θ_2–Computer Networks	{Introduction to Networking, The Network Layer, The Transport and Application Layers, Networking Services, connecting to the Internet, Troubleshooting and the Future of Networking}	LinkedIn – 0.29, Curricular History –0.25, Lattes –0.11

of the "Software Engineering for Web Applications" course contained in the OCW provider. Considering that this book was the best-rated content and that Tom has not accessed this material before, the system instantly retrieves the result.

The entire recommendation process is saved in the knowledge base in order to improve the recommendation process for future recommendations. The recommendation is considered successful if Tom uses the recommendation system next time and the module that was recommended to him appears as a complete module, which indicates that he used the content. If it has not been completed, the engine considers this information to prevent Tom from receiving the same unwanted recommendation again.

The functioning representation of the recommendation system used in the example of use demonstrates the usability, functionality, and relevance of the proposed approach. The system also gains scalability by using topic modeling and retrieval with open source data. We believe that it can be implemented in a real MOOC scenario, where it allows reaching the customized recommendations for each student according to his/her motivations and needs. In addition, it would reinforce the metric effectiveness as a way of identifying a study plan with part of courses from multiple providers.

8. Final Remarks

The use of recommendation systems has been applied with different objectives and with algorithms and processes increasingly optimized as several challenges arise in different domains. This work addresses the use of Linked Open Data and topic modeling and labeling methods to integrate data and create an architecture for a web-based recommendation system capable of recommending courses, modules or parts of courses, and relevant materials of students' interests in multiple MOOCs providers. From this environment, we conceptualize the MOOCs ecosystems formed from such providers, their users, and other actors, describing how this approach can bring benefits to the learning processes, to the platform's sustainability and to the stakeholders. A motivation for this conceptualization is to map roles, actions and interactions between users, allowing the understanding of these platforms as MOOCs and not only as a VLE. Considering the characteristics of the proposed architecture, it is possible to support knowledge reuse within the ecosystem.

We are currently implementing the proposed architecture and preparing the evaluation of the algorithms chosen to recommend items based on experiments *in vitro* and *in vivo*. The first aims at verifying the algorithm efficiency and effectiveness from a controlled experiment. Then the results will be compared with the results from related work. The second is a feasibility study to evaluate the solution with two groups of people who receive a series of pre-established tasks, where the first group performs them using the proposed solution, while the second group does not. For this study, we will invite students from different backgrounds (initial phase, final phase or already working in a specific area), all from a Brazilian higher education institution. At the end of the study, the results will be compared, and the participants will answer a questionnaire, providing some feedback. Finally, we will collect data from documents and repositories used in the study in order to analyze elements of our proposed solution in details.

References

[1]   Charu, C. Aggarwal. 2016. Recommender Systems. Springer International Publishing, Cham, Feb. 2016.

[2]   Sunita, B. Aher and L.M.R.J. Lobo. 2013. Combination of machine learning algorithms for recommendation of courses in E-Learning System based on historical data. Knowledge-Based Systems 51: 1–14.

[3]   Luisa Aires. 2015. E-Learning, online education and open education: A contribution to a theoretical approach. RIED. Revista Iberoamericana de Educación a Distancia 19(1): Sep. 2015.

[4]   Olavo Alexandrino Loiola Pinto Barbosa, Rodrigo Pereira dos Santos, Carina Frota Alves, Claudia Maria Lima Werner and Slinger. Jansen. 2013. A systematic mapping study on software ecosystems from a three-dimensional perspective. pp. 59–81. *In*: Software Ecosystems: Analyzing and Managing Business Networks in the Software Industry. Northampton/USA: Edward Elgar Publishing, 1 Ed.

[5]   Paul Belleflamme and Julien Jacqmin. 2016. An economic appraisal of MOOC platforms: Business models and impacts on higher education. CESifo Economic Studies 62(1): 148–169, Mar. 2016.

[6]   David M. Blei, Andrew Y. Ng and Michael I. Jordan. 2003. Latent dirichlet allocation. Journal of Machine Learning Research 3: 993–1022.

[7]   Paul Bond and Faye Leibowitz. 2013. MOOCs and serials. Serials Review 39(4): 258–260, Dec. 2013.

[8] Fatiha Bousbahi and Henda Chorfi. 2015. MOOC-Rec: A case based recommender system for MOOCs. Procedia—Social and Behavioral Sciences 195: 1813–1822.

[9] Rodrigo Campos, Rodrigo Pereira dos Santos and Jonice Oliveira. 2018. Recommendation systems for knowledge reuse management in MOOCs ecosystems. pp. 46–48. *In*: Anais do XIV Simpósio Brasileiro de Sistemas de Informação (SBSI), editor, XI WTDSI–XI Workshop de Teses e Dissertações em Sistemas de Informação. Caxias do Sul/RS, Brasil. Porto Alegre: SBC.

[10] Paul Cooper. 2017. Data, information, knowledge and wisdom. Anaesthesia & Intensive Care Medicine 18(1): 55–56, Jan. 2017.

[11] Thanasis Daradoumis, Roxana Bassi, Fatos Xhafa and Santi Caballé. 2013. A review on massive e-learning (MOOC) design, delivery and assessment. pp. 208–213. Proceedings 8th International Conference on P2P, Parallel, Grid, Cloud and Internet Computing, 3PGCIC.

[12] Chinmay Dhekne. 2016. MOOCLink: Linking and Maintaining Quality of Data Provided by Various MOOC Providers. Ph.D. thesis, M.S. thesis, Arizona State University.

[13] Tommaso Di Noia, Roberto Mirizzi, Vito Claudio Ostuni, Davide Romito and Markus Zanker. 2012. Linked open data to support content-based recommender systems. Conf. on Semantic Systems.

[14] Derek Greene, Derek O'Callaghan and Pádraig Cunningham. 2014. How Many Topics? Stability analysis for topic models. pp. 498–513. *In*: Joint European Conference on Machine Learning and Knowledge Discovery in Databases. Springer, Apr. 2014.

[15] Benjamin Heitmann and Conor Hayes. 2010. Using linked data to build open, collaborative recommender systems. In 2010 AAAI Spring Symposium Series.

[16] Slinger Jansen, Anthony Finkelstein and Sjaak Brinkkemper. 2009. A sense of community: A research agenda for software ecosystems. pp. 187–190. *In*: 31st International Conference on Software Engineering-Companion Volume.

[17] Yehuda Koren, Robert Bell and Chris Volinsky. 2009. Matrix factorization techniques for recommender systems. Computer 42(8): 30–37.

[18] Thaiana Lima, Rodrigo Pereira dos Santos, Jonice Oliveira and Cláudia Werner. 2016. The importance of socio-technical resources for software ecosystems management. Journal of Innovation in Digital Ecosystems 3(2): 98–113, Dec. 2016.

[19] Davide Magatti, Silvia Calegari, Davide Ciucci and Fabio Stella. 2009. Automatic labeling of topics. pp. 1227–1232. *In*: Ninth International Conference on Intelligent Systems Design and Applications, IEEE.

[20] Dimitrios Malakoudis and Panagiotis Symeonidis. 2016. MoocRec.com: Massive open online courses recommender system. In RecSys Posters, Jul. 2016.

[21] Lynne M. Markus. 2001. Toward a theory of knowledge reuse types of knowledge reuse situations and factors in reuse success. Journal of Management Information Systems 18(1): 57–93.

[22] Jamshaid G. Mohebzada, Guenther Ruhe and Armin Eberlein. 2012. Systematic mapping of recommendation systems for requirements engineering. International Conference on Software and System Process (ICSSP), pp. 200–209.

[23] Diogo Nolasco and Jonice Oliveira. 2016. Detecting knowledge innovation through automatic topic labeling on scholar data. pp. 358–367. *In*: 49th Hawaii International Conference on System Sciences (HICSS), IEEE, Jan. 2016.

[24] Diogo Nolasco and Jonice Oliveira. 2016. Topic modeling and label creation: Identifying themes in semi and unstructured data. pp. 87–112. *In*: Brazilian Symposium on Database-Topics in Data and Information Management. Eduardo Ogasawara, Vaninha Vieira. (Org.), Porto Alegre: SBC.

[25] Christos H. Papadimitriou, Prabhakar Raghavan, Hisao Tamaki and Santosh Vempala. 2000. Latent semantic indexing: A probabilistic analysis. Journal of Computer and System Sciences 61(2): 217–235.

[26] Kuang Shanyun, Shen Qin and Zhou Guolin. 2015. Research on the construction of MOOC learning community ecosystem circle. pp. 199–203. *In*: International Conference of Educational Innovation through Technology (EITT), Vol. 2, IEEE, Oct. 2015.

[27] Pedro Szekely, Craig A. Knoblock, Fengyu Yang, Xuming Zhu, Eleanor E. Fink, Rachel Allen and Georgina Goodlander. 2013. Connecting the smithsonian american art museum to the linked data cloud. pp. 593–607. *In*: Extended Semantic Web Conference, Springer.

[28] Ivo Van Den Berk, Slinger Jansen and Lútzen Luinenburg. 2010. Software ecosystems: a software ecosystem strategy assessment model. pp. 127–134. *In*: Proceedings of the Fourth European Conference on Software Architecture: Companion, ACM.

[29] Ariel Gustavo Zuquello. 2015. OERecommender: um sistema de recomendação de REA para MOOC. Ph.D. thesis, M.S. thesis, State University of Maringá.

Chapter 6

Towards a Computer Vision Based Approach for Developing Algorithms for Soccer Playing Robots

Patrick Hansen, Philip Franco and *Seung-yun Kim**

1. Introduction

The fields of robotics and machine learning intersect in the pursuit of creating artificial intelligence that can dictate the behavior of a robot autonomously. Machine learning deals with learning a function that describes a dataset to make predictions and this can be utilized in robotics to have a robot be able to learn information about its environment and behave accordingly. While machine learning algorithms can be utilized by a robot to learn its environment, an additional algorithm is necessary to control the robot's behavior. This algorithm would be based on the outputs yielded by the machine learning models to account for different states the robot may be faced with.

Within the past decade, there have been numerous publications exploring and applying various areas of machine learning. Our area of interest is computer vision with a focus on utilizing visual stimuli from a robot as inputs to a behavioral model. To address the issue of how to create an artificial intelligence which utilizes machine learning algorithms to drive a controller

Department of Electrical and Computer Engineering, The College of New Jersey, Ewing, NJ 08628, USA.
Emails: hansenp2@tcnj.edu; francop1@tcnj.edu
* Corresponding author: kims@tcnj.edu

algorithm, we will look at the case of soccer playing robots, more specifically *RoboCup*[1] which has two teams of Nao robots playing against each other. Image based machine learning models will be trained to have a robot be able to recognize other robot players (e.g., both teammates and opponents while being able to differentiate between the two) and recognize the soccer ball and predicting the distance to the soccer ball. The machine learning models will yield probabilities describing what the robots see which will be the inputs to a Petri net model that will dictate the robot's behavior. Both the machine learning models and Petri net model will be evaluated to analyze the overall system effectiveness.

1.1 *Related Work*

Petri net (PN) models are finite state automata that can be used to graphically model various antecedent-consequence sequences of actions such as controllers or algorithms [1]. The addition of fuzzy logic or time on the transitions of a PN (i.e., a change in the state of the model) allows for more robust models to be designed. Applying fuzzy set theory to the transitions of a PN leads to a Fuzzy Petri net (FPN) while the addition of time to the transitions leads to a Timed Petri net (TdPN). The addition of either allows for robust models which have been showed can be applied to soccer robot algorithms as well as other game strategies [2, 3]. Furthermore, PN models have been used to model a self-navigating robot through a maze as well as model the optimal path a soccer playing robot should take to score a goal [4, 5].

Using PNs to model soccer playing robots is of interest due to *RoboCup* and the initiative of one day have a team of robots that can compete against humans. This initiative has led to many PN based algorithms designed to control the behavior of humanoid soccer playing robots. In [3], the different robots on a single team were modeled to determine how robots should move to be in the optimal positions to score and in [6] the optimal sequence of actions to complete a team based task were modeled. Similarly, [7] showed that teamwork based actions of soccer playing robots with a focus on passing can be modeled and [8] modeled soccer robot behaviors around having awareness of scoring opportunities.

While PNs have been used to model the behavior or robots, machine learning and computer vision have also been extensively used in modeling the actions of a robot. One major challenge of executing computer vision algorithms on a robot was addressed in [9] is being able to process video in real-time. In [9], the use of deep neural networks was discussed to be applied

[1] RoboCup Standard Platform League, http://spl.robocup.org/.

to real-time object detection by a robot and this concept was applied in [10] for soccer playing robots showing that object detection can be used to identify object necessary to score a goal in a variety of conditions, such as different lighting.

1.2 Proposed Methodology

Our proposed methodology is shown in Figure 1 where we divide the task of creating a model for soccer playing robots into the various sections of this paper. The two primary objectives for the robot are to be able to recognize both other robots and the soccer ball as well as predict the distance to the soccer ball based on the image of the soccer ball. Object detection and distance prediction are two separate machine learning problems each requiring their own models to be trained. The *TensorFlow* object detection API will be used to train a model to classify objects in images seen by the robot as another robot or the soccer ball. Multiple linear regression and multilayer perceptron models will be used to predict the distance from the robot to the soccer ball based on a set of engineered features obtained from the image containing the soccer ball and this will be discussed in Section 3. Section 4 will show the methods used to evaluate both machine learning models.

The last element of the system is the Timed Petri net model which will be described in Section 5. Petri nets allow for controllers to be designed around antecedent-consequence based rules and time can be incorporated to make the state transitions more dynamic. The machine learning models trained in Sections 2 and 3 will yield an initial state for the PN controller based on what the robot sees and the PN will then determine which action the robotic goalkeeper should take.

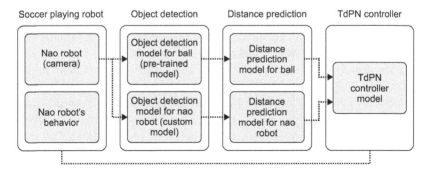

Fig. 1: Block diagram showing overview of proposed methodology. The soccer playing Nao robot will capture images with the cameras in its head, both the soccer ball and other robots will be detected using object detection, and then a model to predict the distance of the two previously mentioned objects will be used. These results will then be used in a Petri net controller to dictate the actions of the Nao robot (in our case we will be focusing on the goalkeeper).

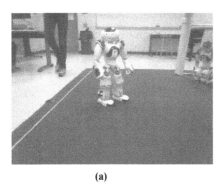 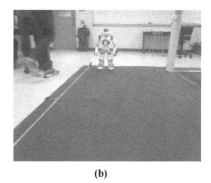

(a)   (b)

Fig. 2: Two example images collected for the object detection model to detect soccer playing Nao robots (these two images are from the testing set and were not trained on).

2. Object Detection Methodology

2.1 Nao Robot Detection

The first objective for a soccer playing robot is being able to detect where other robots are. This requires some form of object detection where the model being used by the robots is trained to recognize other robots. The object detection API provided by *TensorFlow* provides the ability of training a custom model from provided images.[2] Using a custom model is necessary because the target of detecting Nao robots is too niche to find other pre-trained models. One other possibility that may yield sufficient results is detecting the Nao robots as people because they are humanoid. In this case, an existing model trained to detect people could be used instead (e.g., a model trained on the COCO dataset). Both approaches require a dataset of images containing Nao robots and will be explored further in Section 4.

Playing soccer is a real time operation so the objection detection system used needs to be able to execute quickly. Considerations from both the robot and model need to be made to ensure fast enough execution. From the robot, the resolution of the image/video will be deterministic of the time required to process a single frame and identify an object within it. The Nao robots can support a few different resolutions, but the size of 320 × 240 pixels was chosen because it provides a good tradeoff between image fidelity and processing time. The object detection model to be used is *MobileNet* due to its ability to handle real time video [11]. Additionally, the *TensorFlow* object detection API supports this model and the building of custom classifiers using the architecture.

---

[2] TensorFlow Repository, https://github.com/tensorflow/models.

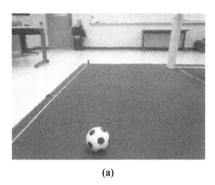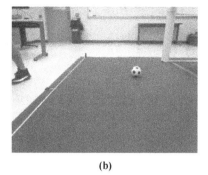

(a)   (b)

Fig. 3: Two images of the soccer ball collected at various distances from the Nao robot for the purpose of distance prediction.

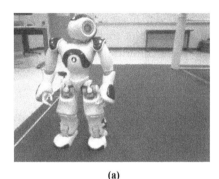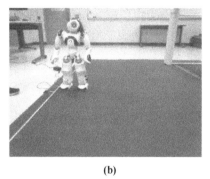

(a)   (b)

Fig. 4: Two images of a Nao robot collected at various distances for the purpose of distance prediction.

2.2 *Image Dataset and Preprocessing*

Regardless of the object detection method, a set of images containing Nao robots is required. In training a custom model, the images collected will be divided into a training and testing subset. The training set is what the model will be fit to and the testing set will be used for validation. To compare to a pre-trained model looking to detect humans, the same testing set can be used for validation to compare the performance of the two models.

In total, 100 images of Nao robots were collected to make up the dataset of which 80% was assigned to the training set and the other 20% to the testing set. The *TensorFlow* object detection API while using the *MobileNet* model outputs boxes around the object detected, so this requires the training data to be set up in this manner. This required the images to be hand labeled with boxes around the object according to the API's documentation.

2.3 Soccer Ball Detection

In addition to locating other robots, the soccer playing robot must be also be able to detect the soccer ball on the field. Soccer balls are a common object and the detection of them have been addressed in datasets that contain common objects. An example of a model that exists and is able to detect a soccer ball is the *MobileNet* trained on the Common Objects in Context (COCO) dataset which is compatible with the *TensorFlow* object detection API. For this reason, a model to detect a soccer ball will not be trained as other models already exist to do so effectively.

3. Distance Prediction Model

The prediction of how far away a soccer ball is from a robot was an issue addressed in [2] where multiple regression was used to make a prediction from engineered regressors. Having robots be able to determine their distance to the soccer ball is paramount to developing a behavioral algorithm. Additionally, a robot being able to determine their distance from another robot is also crucial. This prediction problem will also be addressed with the incorporation of the previously discussed object detection. The machine learning model will also be trained to identify a soccer ball, and if a soccer ball is found the image will be cropped in accordance to the box drawn around the ball to then be used for distance prediction. Features will be engineered to be extracted from the cropped image to then be fed into a machine learning model to predict the distance. We propose using a fully connected neural network with rectified linear unit (ReLU) nonlinearities for this problem and will compare to a multiple regression model as proposed in [2] as a baseline predictor.

3.1 Image Dataset and Distance Labels

For the task of distance prediction, a dataset of images of the soccer ball specified by the *RoboCup* rules were taken. The distances being measured was the length from the robot to the soccer ball in centimeters. Distances from the robot to the soccer ball ranged from 60 cm to 200 cm based on the visibility of the ball from the robot's cameras and the ball was moved in increments of 10 cm because that is the diameter of the soccer ball. Various images were taken at distances from the robot ranging from 60 cm to 200 cm in 10 cm increments to make up the dataset that would be split into training and testing sets. Similarly, an image dataset for pictures of another robot at the same specified distance of 60 cm to 200 cm in 10 cm increments was also taken.

3.2 Feature Engineering

Each image containing the soccer ball and robot was converted to a vector of engineered feature that describes the image. The basis of the feature vector generation is that object detection will be used to locate the soccer ball within the frame and the object detection API will draw a box around where the soccer ball is. From the box drawn around the soccer ball, that piece of the image was cropped and the features were drawn from the cropped portion of the original image only.

Using the cropped portion of the original image containing only the soccer ball, eight features were extracted to describe the distance to the ball in the image. In extracting the features from the images, each image is filtered with a Gaussian blur first and then a Sobel edge detector to find the edges of the soccer ball in each frame. The Sobel filter is a high pass edge detecting filter and the Gaussian blur is a low pass filter to help remove noise around edges before detection [2]. The eight features extracted from each frame are:

1) *Cropped image height*: The height of the cropped portion of the image containing the soccer ball in pixels.

2) *Cropped image width*: The width of the cropped portion of the image containing the soccer ball in pixels.

3) *Number of white pixels*: The number of pixels that are white after passing the image through Sobel filter to find the edges. These white pixels are the edges within the frame which in the cropped frame will be the edges of the soccer ball.

4) *Number of black pixels*: The number of pixels that are black after passing the image through Sobel filter to find the edges.

5) *Number of 0 degree angles*: The number of pixels that have an angle close to 0 degrees after obtaining the gradient of the image from the Sobel filter.

6) *Number of 45 degree angles*: The number of pixels that have an angle close to 45 degrees after obtaining the gradient of the image from the Sobel filter.

7) *Number of 90 degree angles*: The number of pixels that have an angle close to 90 degrees after obtaining the gradient of the image from the Sobel filter.

8) *Horizontal location within image*: The horizontal location of the ball in the original frame encoded as a one-hot distribution signifying if the ball is in the left portion, center portion, or right portion of the image.

These eight features were extracted from each image taken with a soccer ball a known distance away from the robot. This distance serves as the label for each image in this prediction task for which supervised learning is used to fit a model to the obtained data.

3.3 Machine Learning Algorithm

From the images collected, the distance from the robot to the ball was predicted using two different algorithms: multiple linear regression as proposed in [2] and a multilayer perceptron. Multiple linear regression was proposed in [2] due to the linear nature that features describing a soccer ball in an image change with respect to distance. This algorithm will be compared to a multilayer perceptron that uses rectified linear units (ReLU) as the nonlinear activation function. The reason for using a neural network for this prediction problem is to find a better fit for the data because not all the features may be as linear relative to distance as assumed based on the human visual system.

Both these models were trained on the same set of features from Section 3.2 describing the images with the multiple regression model being the baseline for which to compare the neural network model to. In Section 4.1, the training and evaluation methodology for the task of distance prediction is explained and the results are analyzed in Section 4.3.

4. Evaluation of Models

4.1 Robot Detection Evaluation

The *TensorFlow* API, and object detection in general, requires images to be hand labeled commonly with a box drawn around the object of interest in an image. This box is the label from which the model learns where the target object is in the sample images. When building a custom model to detect a new object using the *TensorFlow* API (in our case the target object is a Nao robot), the obtained images need to be hand labeled with boxes drawn around the target object. The labeling of data is a costly process, so as a result the dataset of images collected was relatively small. In total, 100 images were collected.

The chosen metric used to evaluate this model is accuracy. Due to the small dataset size for both training and testing images, accuracy was defined as whether or not a box was drawn around the object that was most likely the Nao robot in the image. Ideally, a larger dataset would have been acquired to better account for the numerous cases that a robot could be presented with during a soccer match, but the purpose was to see if a model could be trained to identify the soccer playing robots which is unique to our problem.

From our test set, the robot in each image was detected as the most likely object in the image to be a robot which validates the idea of using object detection to identify other robots on the playing field. The only issue the

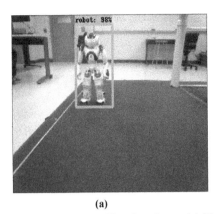

(a)	**(b)**

Fig. 5: Output from object detection model. The model was able to detect the robot in all images as shown in (a) but had issues when the robot was not fully in frame as shown in (b).

model encountered was when the robot was very close causing parts of the robot to move out of frame. While the robot is still detected, the probability that it is a robot is not as high and this is likely due to the majority of the training data containing images where the full robot fits into the frame. Figure 5 shows the results on two test images passed to the object detection model. In the two images, the robot is detected as indicated by the box drawn around the robot and the issue of a close robot is also shown.

4.2 Distance Predictor Evaluation Setup

Due to having a small amount of data due to the need for collecting and hand labeling images of a specific scenario for *RoboCup*, multiple training runs were used for evaluation meaning many models were trained on different training sets from the image dataset. The training and testing split used was 80% of the data for training and the remaining 20% for testing. Images of the soccer ball were taken at a range of 60 cm to 200 cm in increments of 10 cm, so an even distribution of images at each distance was ensured to be present in both the training and testing sets. Comparatively, training these distance predictors takes less time than the robot detector and this time difference is why training multiple models was only done with the distance predictors.

To evaluate the results of both the multiple linear regression and multilayer perceptron, root mean square error (RMSE) was used. RMSE was chosen as the evaluation metric because it measures the deviation of the predictor in the same unit as the predicted variable. Equation 1 defines RMSE:

$$RMSE = \sqrt{\frac{1}{N} \sum_{n=0}^{N-1} (\hat{y}_n - y_n)^2} \tag{1}$$

where $\hat{y}$ is the vector containing the predictions, y is the vector containing the ground truth labels for the distances, and N is the total number of samples (and the length of both vectors $\hat{y}$ and y).

4.3 Evaluation of Distance Prediction Models

Results of the distance predictors are shown in Table 1 where the multiple linear regression model proposed in [2] slightly outperformed the multilayer perceptron for both cases. In general, the RMSE from both models are relatively similar which could be indicative of whether or not the neural network converged to an absolute minimum. Furthermore, the multiple linear regression will be more efficient if implemented on a robot because each prediction only requires one sum of product compared to the many more in the neural network due to the neurons in the hidden layer. For implementation on a robot, lower complexity predictors are desirable due to the embedded processor in the robots which makes multiple linear regression the more desirable model to use. It might be possible for the multilayer perceptron to outperform the multiple linear regression model by increasing the complexity of the neural network but this would also increase the number of arithmetic operations and execution time required for a prediction.

The results shows that distance does have a linear relationship with distance to objects in images and the objects can be common objects like a soccer ball or more complex objects such as the Nao humanoid robot. Additionally, the results suggest that a less complex multiple linear regression model is a more suitable predictor than a neural network and requires far less arithmetic operations making it the more desirable model for implementation on a soccer playing robot.

Table 1: Results of distance predictions models for both the soccer ball and Nao robot (both measured in RMSE).

	Soccer Ball (RMSE in cm)	Nao Robot (RMSE in cm)
Multiple Linear Regression	6.128 ± 0.739	5.195 ± 0.573
Multilayer Perceptron	6.761 ± 1.564	6.487 ± 1.622

5. Petri Net Controller

5.1 Petri Net Models

Petri nets (PN) are mathematical models used to show the flow of data through a system. They are graphical in nature and are able to represent finite state automata in more depth than a state machine (state machines are a subset of PNs). The graphical PN model is represented with nodes called places, transitions, and tokens which represent data moving through the network from place to place through transitions.

Mathematically, the PN model is defined as a 5-tuple:

$$N = (P, T, F, W, M_0) \tag{2}$$

Made up of a finite set of places P, a finite set of transitions T, a finite set of arcs F, a weight function W on the arcs, and an initial marking M_0. In general, a marking M of a PN defines the number of tokens across the places in the network and the initial marking M_0 is the token distribution at the start.

$$P = \{p_1, p_2, ..., p_k\} \tag{3}$$

$$T = \{t_1, t_2, ..., t_m\} \tag{4}$$

$$F = \{f_1, f_2, ..., f_n\} \tag{5}$$

$$M : P \rightarrow N \tag{6}$$

$$W : F \rightarrow N \tag{7}$$

With the mathematical definition of PNs, the graphical model can be defined as places connected by transitions through directed arcs and tokens travel across the places in the network. The places, transitions, and arcs define the behavior of the PN model while the tokens themselves are the data being observed by the network, and the final marking (i.e., where the token or tokens end up) is the result yielded by the network.

Additionally, the arcs connecting the places and transitions are weighted and the individual weights on each arc are specified by the function W. The weight function defines two different types of behaviors depending on whether the arc is directed from a place to a transition or from a transition to a place. Weights on arcs directed from a place to a transition specify the number of tokens that will be consumed by firing the transition which is the number of tokens that need to be present in the place to fire the transition. In contrast, weights on arcs directed from transitions to places specify the number of tokens produced by the transition firing which is the number of tokens that will be present in the place after a transition fires.

Figures 6 and 7 show small example PNs to demonstrate the defined properties. In Figure 6, there are two places p_1 and p_2 connected by transition t_1. Both arcs connecting to t_1 have a weight of one meaning that one token will be consumed from place p_1 and one token will be produced in p_2 when t_1 fires. Transition t_1 is able to fire once the token requirement in p_1 is met. Comparatively, the PN in Figure 7 takes the same network but changes the weights on the two arcs. Now, the arc connecting p_1 to t_1 has a weight of 3 and the arc connecting t_1 to p_2 has a weight of 2. Once p_1 has three tokens in it, t_1 will fire consuming all three tokens and produce 2 tokens in p_2.

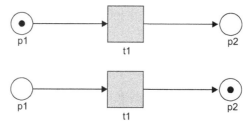

Fig. 6: Example Petri net showing the transition firing characteristic of t1 between places p_1 and p_2.

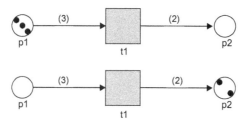

Fig. 7: Example Petri net showing the transition firing characteristic with weights to change to the number of tokens produced and consumed by firing transitions t_1.

Additional rules governing the PN model involve what happens when multiple arcs are directed into a transition or out of a transition. If two places have arcs directed to the same transition, then the token consumption requirement on all connection arcs must be met for the transition to be able to fire. If a transition has more than one arc directed out of it, then the number of tokens specified by the weight of each arc will be produced when the transition fires. Lastly, if a place has multiple arcs connecting to multiple transitions and the token consumption requirements are met for more than one transition to fire, then only one of the transitions is able to fire which can cause ambiguity in the model which will be addressed in Section 5.2 by constraining the PN model to additional mathematical properties governing the firing of transitions.

5.2 Timed Petri Net Models

The Petri net model can be further enhanced through the addition of a time variable on the transitions with a structure called the Timed Petri net (TdPN). Mathematically, TdPN's are defined as the 6-tuple in Equation 8:

$$N_\tau = (P, T, F, W, M_0, \tau) \tag{8}$$

where P, T, F, W, and M_0 are the same as defined in Equations 2–7 for a standard PN and the new variable τ which denotes the time it takes for a transition to fire [3].

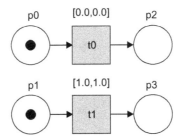

Fig. 8: Example Timed Petri net demonstrating the effect the timing variable τ has on the firing of transitions t_0 and t_1.

The example TdPN shown in Figure 8 demonstrates the addition of the timing variable τ differs this model from a regular PN. In this example, notice how the transition t_0 is red and the transition t_1 is not. Transitions that are red are enabled and are ready for tokens to be fired through them. The time above the transition t_0 [0,0] indicates that a token can fire through it without hesitation, but transition t_1 has a time delay of 1 denoted by [1.0, 1.0]. Transition t_1 will be fired second, after transition t_0 because of this.

5.3 Enhanced Arcs in Petri Net Models

Additional types of arcs exist in PN models to offer different types of behavior for the firing of a transition. In the model that will be introduced in Section 5.4, we employed the use of inhibitor and test arcs. Inhibitor arcs disable transitions if there is a token in the place preceding the arc and are denoted with an empty circle compared to an arrow on regular arcs. Figure 9 demonstrates the effect of an inhibitor arc with there being a token in p_1 which disables the transition t_0. As long as there is a token in p_1, the token in p_0 will not be able to fire through transition t_0 and the resulting Petri net is said to be in a deadlock.

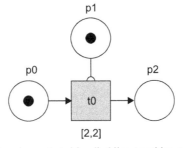

Fig. 9: Effect of Inhibitor Arc demonstrated by disabling transition t_0 as long as there is a token present in place p_1.

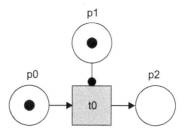

Fig. 10: Places connecting to a transition via a Test Arc (in this case, p_1 connecting to t_0) will not lose their token when the transition fires.

Test arcs are denoted with a filled circle and allow a transition to be enabled if there is a token in the place preceding the arc similarly to how a regular arc functions. The main and only difference between regular and test arcs is the token in the place connecting to a transition via a test arc will not loses its tokens when the transition fires (i.e., test arcs do not consume the tokens when the connecting transition is fired). An example of a test arc is shown in Figure 10 showing that the token will remain in p_1 when the token in p_0 is fired through transition t_0.

5.4 TdPN Controller Design and Implementation

TdPNs are suitable for the modeling of robotic behavior due to their robust modeling capabilities. We will be using the model to develop an algorithm for soccer playing robots with the focus being on the behavior of the goalkeeper. The TdPN will utilize the machine learning models developed in the previous sections in generating the initial marking of the network from which a decision will made based on what it currently sees. Results from the machine learning models that the TdPN will utilize are the locations of the soccer ball and other robots from the object detection and the distances to each from each object detected using the distance predictors.

Figure 11 shows the designed TdPN controller which was designed using the modeling tool TINA.[3] Boxes are drawn on top of the TdPN in Figure 11 so the different components of the model can be more easily differentiated visually. Starting on the left-hand side of the model is the beginning of the Petri net which breaks up an image into 9 sections as denoted by the 9 places. These places are populated with tokens before the PN starts firing as part of the initial marking M_0. Each place in this section corresponds to a section of the image captured by the Nao robot and the tokens in each place are

---

[3] TINA Project Page, http://projects.laas.fr/tina/.

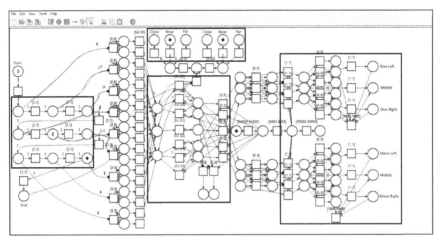

Fig. 11: Timed Petri net (TdPN) controller implemented using the TINA modeling software. The four main components of the model have boxes drawn around them.

determined by whether or not one of the objects of interest are detected in that region of the image via the objection detection models. For the places in this section of the model, 1 and 2 tokens denote the ball and a robot, respectively (these numbers were arbitrarily defined and additional objects can be added by assigning them numbers). Furthermore, the soccer ball takes priority over any other object meaning if an additional object is found in the same region of the image as the soccer ball, the corresponding place in the PN will be marked as only having a soccer ball (i.e., marked with 1 token). In the image shown, this then indicates that the soccer ball is located in the bottom right-hand side of the image and a robot is located in the center of the image. Using this general layout for the TdPN allows for the model to be scaled up through the addition of more places (i.e., more defined regions of the image) and additional objects of interests, such as the goal post or distinct markings on the field.

The initial marking of the TdPN is also defined by the distances to the objects detected within the image. From the collection of images and training of the distance prediction models, the effective range for distance prediction is 60 cm to 200 cm from the robot which is then broken into three categories: close, near, and far. For our purposes, close is defined from 60 cm to 100 cm, near is defined from 100 cm to 160 cm, and far is defined from 160 cm to 200 cm (these ranges can be adjusted as needed based on the time it takes for a robot to move, make a block, or the speed of the ball when kicked by an opposing robot). Depending on the distances to the soccer ball and opposing robot, the goalkeeping robot will either move to the left or right or dive to the left or right. Furthermore, the closer the ball is to the goalkeeper, the more likely the robot is to dive to make a block on the ball.

Within the TdPN, the distances to the objects are accounted for with separate inputs located in the uppermost boxed area of TdPN. This information will be obtained from the distance prediction models and then be classified as close, near, or far according the previously described ranges. In the current TdPN shown in Figure 11, there is a set of two distance inputs; one for the soccer ball and one for an opposing robot. The closer the object is to the goalkeeper, the more likely the goalkeeper is to take an action in defending a shot on goal. Additionally, this information is used in accordance with the whether the objects are in the left, center, or right portion of the image in making a decision. For example, if the ball is close to the goal and on the left side, then the goalie will dive left in contrast to only moving left if the ball is labeled as far.

From the initial marking, which is determined as previously discussed, the resultant decision made by the TdPN is determined by the additional blocked sections in Figure 11. The first block converts the positional tokens for the objects into a form from which a decision can be made. The tokens from this part of the initial marking get converted into tokens distributed across three places. Of these three places, the one with the most tokens will be used in determining the decision of the action the robot will take. The decision made from these three places is determined using a form of a PN comparator to compare the tokens in the three places. Two separate comparators are used to make the decision of the whether the robot should stay in the middle of the goal, move left, move right, dive left, or dive right. Of the six decisions shown on the far right denoted by places, the *stay middle* refer to the same action.

5.5 *TdPN Controller Analysis*

In its current state, the TdPN controller is able make one of five decisions based on the locations and distances to the soccer ball and opposing robots which determines the initial marking of the network. These five decisions resemble those of an actual goalkeeper and the model looks to mimic the decision process that an actual goalkeeper would make when presented with the same information. Without expanding this TdPN (i.e., increasing the size by adding more places and transitions), the current architecture of the model can be modified to recognize more objects than the two it currently can identify. The two objects that the TdPN identifies are the robot and the soccer ball. The position of the soccer ball affects the robots decision twice as much as the position of the opposing robot.

The current architecture of our model allows the model to be scaled up as necessary. Additions to the model that can be made in the future include the ability to segment the image into more pieces rather than the current 9, the ability to add more objects of interests which would also require modifications to the machine learning models, and better generalizing this model to create a new model that can be applied to the rest of robots on the playing field.

6. Conclusion

In this paper, we proposed a computer vision based methodology for developing an algorithm using a Timed Petri net to dictate the behavior of a soccer playing robot, with our focus being on the goalkeeper. We first used object detection models to detect a soccer ball as well as robots in an image that would be captured by a robot. For the detection of robots, a custom model was trained while a pre-trained model for detecting soccer balls was used, but in the future these models would be unified. Furthermore, additional objects of interest, such as the goal or defining marks on the soccer field, could be included in the object detection model. The object detection models draw a box around the detected object in the image, and then the image was cropped to the box containing the object used for distance prediction. For both the soccer ball and other robots, features were first engineered and then a multiple linear regression model was trained on the features to predict the distance to the object in centimeters.

The results from the machine learning models, objection detection, and distance prediction for both the soccer ball and robots were then used as inputs to a Timed Petri net (TdPN) controller used to determine which action the robot goalkeeper should take. Results from the machine learning models were used to determine the initial marking M_0 of the TdPN controller which then runs to determine the action of the robotic goalkeeper should take. In our model, five actions for the robot to take are defined and include stay in the middle of the goal, move left, move right, dive left, or dive right. These actions were chosen to resemble those of an actual goalkeeper. Which action should be taken is based on the distances from the goalkeeper to the ball and opposing robot, and looks to mimic the type of decision an actual goalkeeper would make based on the same information. The designed TdPN controller is also setup in a way that allows for future modifications, such as further segmenting the image or adding more objects of interest which would also require additional machine learning models to be trained.

References

[1] Murata, T. 2013. Petri nets: Properties, analysis and applications. Proc. of the IEEE 77: 541–574.
[2] Hansen, P., P. Franco and S. Kim. 2018. Soccer ball recognition and distance prediction using fuzzy petri nets. IEEE International Conference on Information Reuse and Integration (IRI), Salt Lake City, UT, pp. 315–322.

[3] Franco, P., P. Hansen and S. Kim. 2018. Using timed petri nets to regulate optimal game strategy. International Conference on Computers and Their Applications (CATA), Las Vegas, NV, pp. 207–213.

[4] Kim, S. and Y. Yang. 2018. A self-navigating robot using fuzzy Petri nets. Journal of Robotics and Autonomous Systems by Elsevier 101: 153–165.

[5] Ponsini, D., Y. Yang and S. Kim. 2016. Analysis of soccer robot behaviors using time Petri nets. Proceedings of the IEEE Information Reuse and Integration, pp. 270–274.

[6] Pham, K.T., C. Cantone and S. Kim. 2017. Colored Petri net representation of logical and decisive passing algorithm for humanoid soccer robots. IEEE International Conference on Information Reuse and Integration (IRI), San Diego, CA, pp. 263–269.

[7] Kim, S., D. Ponsini and Y. Yang. 2017. Towards a versatile opportunity awareness algorithms for humanoid soccer robots using time Petri nets. International Journal of Computer Techniques 4(2): 82–94.

[8] Yang, Y., D. Ponsini and S. Kim. 2016. Ball control and position planning algorithms for soccer robots using fuzzy Petri nets. Proceedings of the ISCA International Conference on Computers and Their Applications (CATA), pp. 387–392.

[9] Tenguria, R., S. Parkhedkar, N. Modak, R. Madan and A. Tondwalkar. 2017. Design framework for general purpose object recognition on a robotic platform. International Conference on Communication and Signal Processing (ICCSP), Chennai, pp. 2157–2160.

[10] Susanto, E. Rudiawan, R. Analia, P. Daniel Sutopo and H. Soebakti. 2017. The deep learning development for real-time ball and goal detection of barelang-FC. International Electronics Symposium on Engineering Technology and Applications (IES-ETA), Surabaya, pp. 146–151.

[11] Howard, A.G., M. Zhu, B. Chen, D. Kalenichenko, W. Wang, T. Weyand, M. Adnreetto and H. Adam. 2017. MobileNets: Efficient convolutional neural networks for mobile vision applications. Computing Research Repository (CoRR).

Chapter 7

Context-dependent Reachability Analysis for Hybrid Systems

Stefan Schupp,* Justin Winkens and *Erika Ábrahám*

1. Introduction

Hybrid systems, in which digital controllers interact with a physical, continuous world show increasing presence in various safety-critical applications, e.g., in the automotive sector, in aviation or in automated plants. Along with their increasing usage, also more and more attention is paid to formal methods for their *safety verification*. Many tools have been developed and successfully applied in practice to analyze the behavior of hybrid system models stemming both from academia as well as from industry. Some of these tools are based on theorem proving [15, 26], others on rigorous simulation [4, 17, 33] or satisfiability checking [18, 19, 25, 27]. In this paper we focus on approaches based on *iterative forward reachability* computations [1, 3, 5, 8–10, 13, 21, 22]; for models in which the evolution of the quantities over time follows non-linear functions, these methods are also known as *flowpipe-construction-based* techniques.

Hybrid systems often posses complex behavior involving numerous physical quantities with different temporal dynamics. Interesting recent developments in hybrid system verification addressed improving scalability to be able to handle higher-dimensional systems, i.e., systems which involve large numbers of physical quantities, for example using *decomposition* to reduce complex problems to several smaller problems [7, 14, 31]. Approaches

RWTH Aachen University.
* Corresponding author: stefan.schupp@cs.rwth-aachen.de

and tools have been demonstrated to be able to handle up to hundreds of variables simultaneously on some case studies. However, despite impressive developments, available tools still struggle with the verification of practically relevant complex hybrid systems.

Typically, these tools specialize on a certain verification technique for a certain subclass of hybrid systems, e.g., timed systems [5, 9, 10], linear hybrid systems [1, 21, 27] or non-linear hybrid systems [13, 25]. our previous work [31] we presented an approach based on decomposition, which divides the state space of a hybrid system into sub-spaces whose behavior is piecewise independent from each other.

In this work we presented an extension to our previous approach which (i) automatically finds the finest decomposition of an input system, and (ii) analyzes the different components and dynamically selects dedicated methods for their analysis based on information about the dynamics in that specific sub-space.

This paper extends our work [32] by generalizing some formalisms and extending our method by specific approaches to handle model parts within the expressivity of rectangular automata:

- we develop data structures for *rectangular automata*, a subclass of hybrid automata that are more expressive than timed automata but for which there are more efficient reachability analysis methods,
- we implement tailored methods for the one-step reachability analysis of rectangular automata, and
- we extend our dynamic decomposition methods to detect and handle subspaces with rectangular dynamics automatically.

All of our extensions are implemented and publicly available in our C++ library HyPro [30].

The remaining contents are organized as follows: In Section 2 we provide some preliminaries before we explain our approach to reduce the computational effort for reachability analysis in Section 3. Section 4 provides experimental results and Section 5 concludes the paper.

2. Hybrid Systems Safety Verification

Let $\mathbb{R}$ denote the set of all real numbers and $\mathbb{R}_{\geq 0}$ the non-negative reals. For a finite ordered set $X = \{x_1, \ldots, x_d\}$ of real-valued variables we define $\dot{X} = \{\dot{x}_1, \ldots, \dot{x}_d\}$ and $X' = \{x'_1, \ldots, x'_d\}$. Let $Pred_X$ denote a set of *predicates* over X (in our applications these will be conjunctions of linear real-arithmetic constraints over X). For a predicate $\varphi \in Pred_X$ and a (*variable*) *valuation* $v = (v_1, \ldots, v_d) \in \mathbb{R}^d$, by $v \models \varphi$ we denote that replacing all free occurrences of variables x_i in φ by v_i evaluates φ to *true*. The meaning of $(v, v') \models \varphi$ for $\varphi \in Pred_{X \cup X'}$ is defined similarly, replacing each x_i by v_i and each x_i' by v_i'. The definition of $(v, \dot{v}) \models \varphi$ for $\varphi \in Pred_{X \cup \dot{X}}$ is analogous.

2.1 Hybrid Automata

In order to verify hybrid systems, we need to provide formal models for them. Among others, hybrid systems are often modeled by *hybrid automata*.

Following the definition from [23], a *hybrid automaton* (*HA*) is a tuple $\mathcal{H} = (Loc, Var, Flow, Inv, Edge, Init)$ with the following components:

- A finite set *Loc* of *locations* or *control modes*.
- A finite ordered set $Var = \{x_1, \ldots, x_d\}$ of real-valued *variables*; we also use the notation $x = (x_1, \ldots, x_d)$ and call d the *dimension* of $\mathcal{H}$.
- *Flow* : $Loc \rightarrow Pred_{Var \cup \dot{Var}}$ specifies for each location its *flow* or *dynamics*.
- *Inv* : $Loc \rightarrow Pred_{Var}$ assigns to each location an *invariant*.
- *Edge* $\subseteq Loc \times Pred_{Var} \times Pred_{Var \cup Var'} \times Loc$ is a finite set of *discrete transitions* or *jumps*. For a jump $(l_1, g, r, l_2) \in Edge$, l_1 is its *source* location, l_2 is its *target* location, g specifies the jump's *guard* and r its *reset*.
- *Init* : $Loc \rightarrow Pred_{Var}$ assigns to each location an *initial* predicate.

Let $\mathcal{H} = (Loc, Var, Flow, Inv, Edge, Init)$ be a hybrid automaton with dimension d and variables $x = \{x_1, \ldots, x_d\}$. A *state* of $\mathcal{H}$ is a pair $\sigma = (l, v) \in Loc \times \mathbb{R}^d$. A *symbolic state* (l, φ) consists of a location $l \in Loc$ and a predicate $\varphi \in Pred_{Var}$ and represents the state set $\{(l, v) \in Loc \times \mathbb{R}^d \mid v \models \varphi\}$.

A sequence $(l_0, v_0), \ldots, (l_n, v_n)$ of states of $\mathcal{H}$ is a (finite) *run* of $\mathcal{H}$ if $v_0 \models Init(l_0)$, $v_i \models Inv(l_i)$ for each $i = 0, \ldots, n$, and for each $i = 0, \ldots, n - 1$ one of the following two conditions holds:

- *Flow*: $l_i = l_{i+1}$ and there exist $\delta \in \mathbb{R}_{\geq 0}$ and a continuous, over $(0, \delta)$ differentiable $f : [0, \delta] \rightarrow \mathbb{R}^d$ such that $v_i = f(0)$, $v_{i+1} = f(\delta)$ and for all $0 < \delta' < \delta$ we have $(f(\delta'), \frac{df}{dt}(\delta')) \models Inv(l_i) \wedge Flow(l_i)$.
- *Jump*: There exists $e = (l_i, g, r, l_{i+1}) \in Edge$ with $v_i \models g$ and $(v_i, v_{i+1}) \models r$.

A sequence $(l_0, \varphi_0), \ldots, (l_n, \varphi_n)$ of symbolic states of $\mathcal{H}$ is called a *symbolic run* of $\mathcal{H}$ and represents the set of all runs $(l_0, v_0), \ldots, (l_n, v_n)$ of $\mathcal{H}$ for which $v_i \models \varphi_i$ for each $0 \leq i \leq n$.

A state is *reachable* if there exists a run leading to it. The *reachability problem* poses the question whether a given state set contains any states reachable in a given hybrid automaton.

Based on the type of predicates in their definitions, we can define different subclasses of HA (see Table 1). For instance, *timed automata* (TA) allow only

Table 1: Decidability results for subclasses of hybrid automata, defined by conjunctions of the respective types of predicates (TA = timed automata, IRA = initialised rectangular automata, RA = rectangular automata, LHA I = hybrid automata with constant derivatives; LHA II = hybrid automata with linear ODEs; HA = general hybrid automata; c: rational constant; e, e', $\dot{e}$: arithmetic expressions over *Var*, *Var* ∪ *Var'* resp. *Var* ∪ *Var'*; e_{lin}: linear arithmetic expression over *Var*; ~ ∈ {<, ≤, =, ≥, >}).

Subclasses	Flows	Invariants Guards	Resets	Bnd. Reach.	Unbnd. Reach.
TA	$\dot{x}_i = 1$	$x_i \sim c$	$x'_i = 0, x'_i = x_i$	✓	✓
IRA	$\dot{x}_i \sim c$	$x_i \sim c$	$x'_i \sim c, x'_i = x_i$	✓	✓
RA	$\dot{x}_i \sim c$	$x_i \sim c$	$x'_i \sim c, x'_i = x_i$	✓	X
LHA I	$\dot{x}_i = c$	$e_{lin} \sim 0$	$x'_i \sim e_{lin}$	✓	X
LHA II	$\dot{x}_i = e_{lin}$	$e_{lin} \sim 0$	$x'_i \sim e_{lin}$	X	X
HA	$\dot{e} \sim 0$	$e \sim 0$	$e' \sim 0$	X	X

variables that are *clocks* with derivative 1, invariants and guards that are conjunctions of constraints comparing clock values to constants, and resets that either set clock values to 0 or leave them unchanged; the reachability problem for TA is decidable (PSPACE-complete). *Rectangular automata* (RA) extend the expressivity of timed automata by allowing constant derivatives and non-deterministic resets from *rectangular sets*, which are cross products of intervals with rational or infinite bounds. The reachability problem for *initialised* rectangular automata (IRA) can be reduced to that of TA [24], where initialised means intuitively that if the dynamics of a variable changes by taking a jump then the jump resets the value of the variable to a (non-deterministically choosen) constant. Consequently, decidability results for IRA are as for TA while in general the unbounded reachability problem is not decidable for RA. If we allow constant derivatives and linear expressions in flows, invariants, guards and resets, reachability via runs with a bounded number of jumps is still decidable. For dynamics described by (linear or non-linear) ordinary differential equations (ODEs) even bounded reachability is undecidable.

2.2 *Forward Reachability Analysis*

Timed automata

The reachable states of a timed automaton can be computed as a finite union of *zones*, which are state sets that can be represented by symbolic states whose predicates are conjunctions of constraints of the form $x_i \sim c$ or $x_i - x_j \sim c$ with ~ ∈ {<, ≤, =, ≥, >} and $c \in \mathbb{Q}$. Zones are defined by special types of convex polytopes in $\mathbb{R}^d$ (see Figure 1). Based on the restricted form of the defining constraints, *difference bound matrices* (*DBM*) [6, 16] offer an

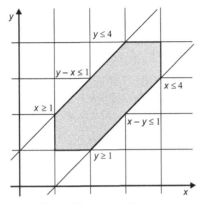

Fig. 1: Illustration of a zone.

efficient representation for zones. For example, the zone in Figure 1 can be represented by a DBM

$$
D = \begin{array}{c} \\ \mathbf{0} \\ x \\ y \end{array}
\begin{array}{ccc}
\mathbf{0} & x & y \\
\left(\begin{array}{ccc}
(0, \leq) & (-1, \leq) & (-1, \leq) \\
(4, \leq) & (0, \leq) & (1, \leq) \\
(4, \leq) & (1, \leq) & (0, \leq)
\end{array} \right)
\end{array}
$$

Each constraint $x_i - x_j \sim c$ is represented by an entry $D_{i,j} = (c, \sim)$ in the DBM where an auxiliary dimension $\mathbf{0}$ with constant zero value has been introduced to allow a normalized representation $x_i - \mathbf{0} \sim c$ of constraints $x_i \sim c$. Thus for a set of n clocks a DBM of size $(n + 1) \times (n + 1)$ is required to represent a zone.

To compute the set of reachable states of a timed automaton, flow and jump successors of the initial state set represented by DBMs can be computed in an alternating fashion. To compute flow successors of a given zone in a given location of a timed automaton, we increase all upper bounds in the entries $D_{i,0}$ for each clock x_i to the largest value still allowed by the invariant (which might be $+\infty$). Similarly for discrete jumps, intersections with guards as well as clock resets can be represented by adjusting the DBM entries. For further details about timed automata model checking we refer to [2].

Rectangular automata

Assume a rectangular automaton and an initial state set represented symbolically by a location and a conjunction of linear constraints over the variables of the automaton. Starting from this initial state set, the reachable states of a rectangular automaton can be described by a (possibly infinite) union of symbolic states (l, φ) whose predicates are conjunctions of linear constraints over the variables of the automaton. Flow successors of states represented symbolically by (l, φ) can be computed by expressing them via quantified linear-arithmetic formulas with free variables from *Var*:

$$\exists t.\ \exists x_{pre}.\ t \geq 0 \wedge \varphi[x_{pre}/x] \wedge Flow\ (l)[x_{pre},\ x/x,\ x'] \wedge Inv\ (l)$$

For example the set of reachable states constructed in Figure 2 with initial condition $1 \leq x \leq 3$ and flow $1 \leq \dot{x} \leq 2$ can be described as follows (for simplicity we assume a trivially true invariant in this example):

$$\exists t.\ \exists x_{pre}. t \geq 0 \wedge 1 \leq x_{pre} \wedge x_{pre} \leq 3 \wedge t + x_{pre} \leq x \wedge x \leq 2t + x_{pre}.$$

Using quantifier elimination techniques, e.g., Fourier-Motzkin variable elimination, we can eliminate quantifiers in order to simplify the formulas. For the above example quantifier elimination shows that by letting time progress, x can take all values larger or equal 1:

$$\exists t. \exists x_{pre}. t \geq 0 \wedge 1 \leq x_{pre} \wedge x_{pre} \leq 3 \wedge t + x_{pre} \leq x \wedge x \leq 2t + x_{pre}.$$
$$\Leftrightarrow \exists t. t \geq 0 \wedge 1 \leq 3 \wedge x - 2t \leq 3 \wedge 1 \leq x - t \wedge x - 2t \leq x - t$$
$$\Leftrightarrow 1 \leq x \wedge -1 \leq x$$
$$\Leftrightarrow 1 \leq x$$

A similar approach can be used to compute successor states of discrete jumps. For further details we refer to [11].

Hybrid automata with constant derivatives and linear constraints

For variables with piecewise constant derivatives and linear constraints (LHA I), the states reachable by flows and a bounded number of jumps can be represented either symbolically similarly as for rectangular automata

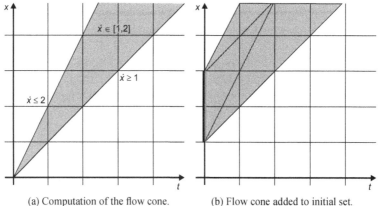

(a) Computation of the flow cone.   (b) Flow cone added to initial set.

Fig. 2: Illustration of the computation of the time successor states in a location of a rectangular automaton for the variable x. Initially $x \in [1, 3]$ and $\dot{x} = [1, 2]$.

above or by the representations mentioned below for LHA II. In the former case, the iterative reachability computations apply quite analogously, but the conditions can now be specified by any set of linear constraints, not only by constant bounds on variables. In the latter case, the computations are similar to the ones for LHA II described below, but we do not need segmentation as the behavior is linear thus segmentation does not increase the precision.

Hybrid automata with linear ODEs and linear constraints

For the more general case of hybrid automata with linear ODEs (LHA II), methods based on *flowpipe construction* can be used. Similar to the method mentioned above for rectangular automata, flowpipe-construction-based methods apply iterative forward reachability computations, starting from some initial state set and alternatingly over-approximating flow and jump successors. However, as the behaviour (according to the solutions of linear ODEs) is now in general non-linear, efficient approaches compute a set of *convex linear sets*, whose union *over-approximates* the states reachable from an initial state set by a *bounded* number of jumps (*jump depth*) and a bounded time ellapse between two successive jumps (*time horizon*). In some cases, when a fixed-point can be detected during analysis, these approaches are also able to make statements about unbounded reachability, but in general this is not the case.

The model class LHA II contains those hybrid automata, whose dynamics in each location l can be specified by a system of linear ODEs over the model's variables $x = (x_1, \ldots, x_d)$:

$$\dot{x} = Ax.$$

Starting from some initial variable values $x_0 = (x_{0,1}, \ldots, x_{0,d})$, according to the above dynamics, after t time units the variables will reach the values $x(t) = e^{tA} \cdot x_0$, where e^{tA} is the matrix exponential for tA. In order to compute bounded reachability, even if we start from a single state, due to non-determinism and non-linear behaviour in the iterative successor computations we need to extend the above solution to the initial value problem to handle initial state *sets* X_0, and to *over-approximate* the set of all states reachable within given time *intervals*.

To over-approximate a *flowpipe*, formed by the trajectories of time evolutions from a set of initial states within a time interval $[0, T]$, i.e., to over-approximate the set of all states reachable *within* time T from a set X_0, standard techniques discretize $[0, T]$ into *time segments* $[0, \delta], \ldots, [(N-1)\delta, N\delta]$ of size $\delta = \dfrac{T}{N}$ and compute *flowpipe segments* Ω_i that over-approximate all states reachable from X_0 within time $[i\delta, (i+1)\delta]$. The computation of the first segment Ω_0 which safely over-approximates reachability within the time interval $[0, \delta]$ is more involved, while all following segments $\Omega_1, \ldots, \Omega_{N-1}$ can be obtained by linear transformations

$$\Omega_{i+1} = e^{\delta A} \cdot \Omega_i.$$

To account for discrete jumps, we compute for each flowpipe segment Ω_i its intersection with all outgoing jumps' guards and apply the corresponding effects to the intersections. The obtained sets are considered as initial sets in the target locations in which again flow successors are computed. To reduce the growth of the search tree, jump successor sets can be *clustered* or *aggregated* into a fewer number of successor sets per target location. Figure 3 shows the algorithm for the computation of flow and jump successors for a given initial valuation set in a certain initial location. Figure 4 illustrates the reachability computation graphically on an example.

The datatypes for the representation of the flowpipe segments Ω_i play an important role in verification, as they strongly affect both precision and efficiency. Popular representations use geometric objects (boxes, polytopes, zonotopes, etc.) or other symbolic representations (support functions, Taylor models, etc.). Each of these representations comes with individual advantages and disadvantages regarding memory requirements and the complexity and precision of the operations that are needed in the reachability analysis (linear transformation, intersection, union, Minkowski sum, etc.). For example, boxes are amongst the fastest set representations but they introduce large over-approximation errors. In contrast to that, support functions provide arbitrary precision but the operations required to obtain this precision are more involved. Convex polytopes are amongst the most precise state set representations but at the same time the computation with them might become expensive with the increasing of the state space dimension.

procedure COMPUTEFLOWPIPE(l, R_0, δ, n)
    $cur \leftarrow R_0$ ▷ state set to compute time successors for
    $F \leftarrow \{R_0\}$ ▷ set of flowpipe segments
    $J \leftarrow \emptyset$ ▷ set of jump successor sets
    while !terminate **do**   ▷ while time horizon not reached

        $cur \leftarrow$ FLOWSUCC(cur, δ)   ▷ compute next flowpipe
    segment
        $F \leftarrow F \cup cur$
        $J \leftarrow J \cup$ JUMPSUCC(cur)   ▷ jump successors of current
    segment
        return (F, J) ▷ return flowpipe segments and their jump
    successors

Fig. 3: Algorithm to over-approximate the flowpipe from a valuation set R_0 in location l using time step length δ and n segments.

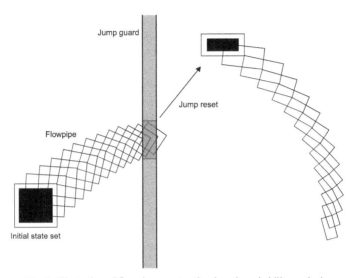

Fig. 4: Illustration of flowpipe-construction-based reachability analysis.

Hybrid automata with non-linear ODEs

For this class of models more involved computations are required to over-approximate flow and jump successors. In this paper we do not cover this case. For further reading we refer to [12].

2.3 *Variable Set Separation*

The applicability of the above reachability analysis algorithms strongly depends on the concrete models, especially on the number of their variables. To reduce the effort for reachability computations, in our previous work [31] we investigated separating *syntactically independent* subsets of variables of a given hybrid automaton $\mathcal{H}$: We seek for a partitioning of the variable set $Var = V_0 \cup \ldots \cup V_k$ of $\mathcal{H}$ into disjoint subsets V_i such that all guards and invariants $\varphi \in Pred_{Var}$ in $\mathcal{H}$ are decomposable into predicates $\varphi = \varphi_0 \wedge \ldots \wedge \varphi_k$ where $\varphi_i \in Pred_{V_i}$; we have similar requirements for flows and resets. If all these criteria are met, we call $D = V_0 \cup \ldots \cup V_k$ a *syntactically independent decomposition* of the variable set of $\mathcal{H}$.

Syntactically independent decompositions allow for a *compositional* reachability analysis where each variable partition is treated individually in a corresponding *sub-space*. In [31] we used this property to effectively reduce the state space dimension for the reachability computations for LHA II. To account for implicit time synchronization, we established connections between flowpipe segments in different sub-spaces that were computed for the same time segments. Based on this implicit time synchronization, predicates

$\varphi \in Pred_{Var}$ in $\mathcal{H}$ such as invariants or guards are only satisfied for a time segment if all decomposed predicates φ_i are satisfied during the same time segment in all sub-spaces. On the one hand, this decomposition speeds up the search remarkably, but on the other hand it introduces some additional over-approximation errors: If a flowpipe segment in a sub-space has a non-empty intersection, e.g., with a jump guard but it is not fully contained in it then we do not exactly know at which time points the guard is true in the corresponding time segment, therefore we need to consider the full flowpipe segments in the other sub-spaces.

3. Context-based Reachability Analysis

3.1 Variable Set Separation Revisited

In the variable set decomposition approach [31] we aimed at grouping variables into three classes: (1) A set of discrete variables that are syntactically independent from all other variables not in the set. (2) A set of variables with constant derivatives that are syntactically independent from all other variables not in the set. (3) All the remaining variables. We required this decomposition to be fixed manually by the user for the whole automaton $\mathcal{H}$.

In this work, we aim at providing an *automated* approach for finding maximal syntactically independent variable sets. To achieve this, we compute a dependency graph $\mathcal{G} = (V, E)$, whose nodes $V = Var_{\mathcal{H}}$ represent the variables and whose edges $(x_i, x_j) \in E$ represent syntactic dependence between the variables x_i and x_j in any location's flow or invariant or in a jump's guard or reset. For instance, when the evolution of x_i syntactically depends on the value of x_j or vice versa. The connected components of $\mathcal{G}$ provide the finest syntactically independent variable set decomposition.

Note that set union preserves syntactical independence, therefore if the finest decomposition contains too many partitions, several partitions can be united to achieve a coarser decomposition.

3.2 Context-sensitive Reachability Analysis

In many applications, for instance when digital controllers are part of the model, the variables of a given hybrid automaton can be classified into continuous physical quantities and discrete variables representing the state of the control program. Additionally the continuous variables can be classified into several subclasses based on the nature of their dynamics and the shape of conditions and reset functions. This classification basically reflects the different subclasses of hybrid automata, as presented in Table 1, i.e., we can separate variables that behave as clocks of a timed automaton, or variables with constant derivatives, variables with dynamics from intervals or general linear or non-linear ODEs.

For the reachability analysis, the time evolution for discrete variables does not need to be computed, as the values of discrete variables remain unchanged during time evolution. Furthermore if we can detect a certain subclass of variables, we can make use of specialized approaches.

For instance, if only clocks are involved, we can exploit this by performing reachability computations which are based on DBM representations in one step instead of the time-discretizing approach which is used for general LHA II. Analogously we can make use of specialized approaches if the automaton is a rectangular automaton and compute the set of reachable states by means of linear predicates.

Thus, in our approach we try to *customize* reachability computations according to the different dynamics in syntactically independent variable sets to further increase the efficiency. In [31] we applied special reachability computation techniques only to syntactically independent discrete variables (in the first set). In this work we generalize our technique to better exploit the individual variable dynamics in different variable partitions by using tailored approaches which are available for certain subclasses.

We introduce the following classification of the variables according to their dynamics and available analysis methods:[1] (1) *discrete class* (zero derivative), (2) *timed class* (derivative 1), (3) *constant class* (constant derivative), (4) *rectangular class* (derivative from rectangular set) and (5) *linear class* (linear ODE). In this work we do not consider non-linear ODEs, mainly due to implementation issues. Note that most classes contain all other classes with lower indices (e.g., (3) contains both (1) and (2)), but there are exceptions ((2) is disjoint from (1) and (4) is disjoint from (5)). For a variable partitioning, to each partition we assign a *context* which is the class with the smallest index $1 \leq i \leq 5$ such that the dynamics of each variable from the partition falls into that class in each location. For example, if a partition contains discrete variables and clocks then its context is the constant class.

We can assign to each context a customized reachability analysis technique.

(5) In our setup the linear class is the most general and therefore default context, which requires the usage of classical flowpipe-construction-based approaches as presented in Section 2.2 for LHA II.

(4) The rectangular class requires an approach different from all the other approaches (see Section 2), i.e., this class is not directly contained in any other class and thus there is no more general approach available.

---

[1] Rectangular automata are a subclass of LHA II but the utilized analysis methods do not reflect this relation.

(3) For the constant class we can apply any method for LHA I to compute flow successors without flowpipe segmentation in one step for unbounded time duration. Note that the analysis method for the rectangular class is a generalization of this approach and thus can also be used.

(2) If we can syntactically separate a set of variables within the timed class then we can use DBM-based analysis methods suitable for timed automata which are far more effective and precise.

(1) Finally, for the discrete class the flowpipe computation reduces to identifying jump successors.

3.3 Implementation Details

In previous works [28, 29] we have introduced the concept of *tasks* and *workers* in hybrid systems reachability analysis. A task $t = (l, R, T)$ stores the information that a flowpipe computation needs to be carried out in a location l for initial valuation set R and time duration $[0, T]$. Tasks are stored in queues and executed by workers, where each worker is implemented as a separate thread. During these executions, each worker might create follow-up tasks for jump successors, which can be distributed over shared global queues to other workers.

We have extended this concept by introducing *context-sensitive workers* to exploit dedicated reachability analysis methods in different contexts and to further improve the scalability of our method. In the following we extend the concept of a symbolic state (l, R) to sub-spaces and use $(l, R_1, \ldots, R_n)$ to denote decomposed sets with R_i being the projection of R to the sub-space of V_i. Thus in the context of a decomposition in n syntactically independent variable sets our tasks are of the form $t = (l, R_0, \ldots, R_n, T)$. Note that the cross product of the projections over-approximates the initial state set (assuming the same order of the dimensions).

After reading the input automaton $\mathcal{H}$, a syntactically independent decomposition $D_{\mathcal{H}} = V_0 \cup \ldots \cup V_n$ is computed using the graph-based approach as presented before. Based on this decomposition, the initial valuation set $Init_{\mathcal{H}}(l)$ for each location l is projected onto the sub-spaces spanned by the variable sets V_i in l, resulting in the valuation sets $Init_i(l)$. For each location, its initial task $(l, Init_1(l), \ldots, Init_n(l), T)$ (where T is the time horizon) is pushed into the working queue and can be grabbed by a worker, which will execute all needed successor computations in the sub-spaces using time synchronization between them, and potentially push jump successors back into the working queue.

As mentioned in Section 2.3, we require implicit synchronization between the different contexts to check conditions such as invariants, guards or intersections with bad states. Synchronization via time intervals between

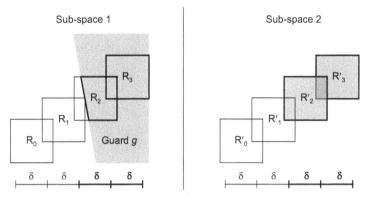

Fig. 5: Synchronization over time intervals for flowpipe segments R_2, R_3 in sub-space 1 maps to segments R'_2, R'_3 in sub-space two.

different sub-spaces may introduce over-approximation errors. We illustrate this using the synchronization on discrete jumps (see Figure 5). In the linear context (class (5)) it is impossible to get the exact time interval T_g when a guard predicate g for a discrete jump is satisfied. The time intervals associated with each segment however allow us to obtain an over-approximation $T'_g \supseteq Tg$ (in Figure 5 the union of the time segments of R_2 and R_3). Using T'_g for synchronization, we identify the corresponding flowpipe segments in all other sub-spaces (R'_2 and R'_3 in Figure 5, right). This introduces additional over-approximation errors, as the exact time interval T_g when the predicate g is satisfied cannot be obtained and used for synchronization between the sub-spaces.

Even though the decomposition of the state space into sub-spaces indicates that the analysis of those can be handled independently, collecting and distributing information over the different sub-spaces is beneficial. As predicates such as guards and invariants need to be satisfied in all sub-spaces at the same time, we can avoid unnecessary checks: For instance knowing that a predicate, e.g., a guard is not satisfied in a certain sub-space during a certain time interval directly allows to skip the check for this guard for the same time interval in all other sub-spaces.

We can further exploit this fact by observing that intersection computation in different contexts comes with different efforts and different consequences. For instance, if a guard g of a jump is not enabled in a discrete context then we know for sure that this transition will not be enabled for the whole flowpipe and can skip consecutive checks in all other sub-spaces during the whole computation of the flowpipe. As another example, it is precise and computationally cheap to gain timing information from a timed or a rectangular context, which can be transferred to the according predicates in other sub-spaces in order to avoid expensive operations.

To this end, verification in the discrete context, followed by the timed and rectangular context and finally the linear context allows for a maximal information extraction and distribution with minimal cost during running time.

Handler-based Contexts

To reduce the implementation overhead and avoid duplicate code, our context-based workers may be assembled modularly from pre-defined handlers. The general approach for reachability analysis as presented in Figure 3 can be divided into several subtasks which can be handled individually. Abstracting those subtasks, for instance computing the intersection with the invariant or computing a jump successor allows to further modularize reachability computation. While a general approach towards this was presented in [20], we aim at refining this idea. As already stated, the basic flowpipe construction-based reachability analysis method distinguishes between computation of time- and jump-successor states. The computation of time successor states can further be partitioned into the computation of the time successor state set (l, R), the validation of the invariant $(l, R') = (l, R \cap Inv (l))$, the intersection with bad states $(R' \cap R_{bad} = \emptyset$?) and testing whether outgoing transitions $(l, g_i, r_i, l') \in Edge$ from the current location are enabled $(R' \cap g_i = \emptyset$?). Jump-successor computation involves application of the reset function r_i and intersection with the target location's invariant. Note that further post- and pre-processing steps for both, the time—as well as the jump-successor computation can be applied or even may be neccessary, depending on the type of dynamics. For instance fixed-point tests can be included as a post-processing of the jump-successor computation, while for example aggregation and clustering can be seen as potential pre-processing steps and make sense when computing LHA II reachability.

While some of those steps vary depending on the context, some are similar in multiple subspaces. The intersection with the invariant condition, as an example is similar in subspace classes (1)–(3) and (5) and thus can be computed by the same code.

To exploit this property and allow for a dynamic creation of contexts, to reduce duplicate code as well as to ease the extension of existing approaches we introduce handler-based contexts. While a general template of a worker is provided, a central *decision entity* allows to implement creation of handlers based on the current context with a fall-back to the most general approach (all classes except (4) fall back to (5)). In the current setup we provide handlers for each class for

- the creation of the first flowpipe segment (required for LHA II),
- the computation for time successors,

- the intersection of state sets with invariants,
- the test for emptiness of a state set with the set of bad states,
- the test for emptiness of a state set with a guard of an outgoing transition,
- the application of a reset function to a state set and
- the creation of follow-up tasks from a discrete jump.

Note that the general context template allows to instantiate pre- and post handlers for each of those mentioned handlers to allow for specialized pre- and post-processing.

The global *decision entity*, which is aware of the number and types of subspaces may decide during running time, which handlers to instantiate for which subspace.

4. Experimental Results

To test our approach, we have conducted several experiments using our tool prototype HyDRA, which is based on HyPro [30], a C++ library providing modules for the development of flowpipe-construction-based reachability analysis methods for linear hybrid systems (LHA II).

We have used a set of commonly known benchmarks, including the bouncing ball (bball), an instance of Fisher's mutual exclusion protocol (fisher), the model of a vehicle platoon (platoon), the simplified model of a temperature control of a reactor (rods), an artificial 5D linear switching system (sw5), and a model of two leaking tanks with a controlled inflow (2tanks). All experiments were carried on an Intel Core i7 (4 × 4 GHz) CPU with 16 GB RAM. The timeout was set to 10 minutes and we used a memory limit (MO) of 8 GB. The running times for our experiments can be found in Table 2. To test our extension towards rectangular automata and decomposition thereof we created a small toy example which contains both, linear dynamics

Table 2: Running times (in seconds) for a selection of benchmarks with context-sensitive workers (sep.) and without using boxes (box) and support functions (sf) with different time step sizes. Timeouts (TO), memouts (MO) and unsuccessful verifications (†) are marked.

benchmark		box $\delta = .1$		box $\delta = .01$		sf $\delta = .01$		sf $\delta = .001$	
		no sep.	sep.	no sep.	sep.	no sep.	sep.	no sep.	sep.
non-dec.	bball	†	†	**0.1**	**0.1**	**0.14**	0.15	**0.51**	0.54
	sw5	†	†	†	†	†	†	**0.32**	**0.32**
dec.	fisher	**5.8**	8.25	**5.14**	74.9	TO	285	TO	TO
	platoon	†	†	†	†	5.17	**4.6**	19.8	**15.5**
	rods	**0.13**	**0.13**	**0.39**	0.43	9.98	**4.13**	TO	339
	2tanks	1.16	**0.71**	**0.75**	0.83	TO	**1.22**	TO	8.78

and rectangular dynamics. Due to the lack of published benchmarks for rectangular automata we used an additional artificial model with 5 variables (5variable_system) taken from [11]. Furthermore, we created an equivalent instance of fisher using a rectangular automaton model to test our approach, as the original dynamics are constant. The running times for those experiments can be found in Table 3.

In our experiments we varied the state set representation between boxes (box) and support functions (sf), and the time step size δ between 0.01 and 0.001 for the analysis of LHA II and the decomposition thereof. The configurations denoted by sep. denote runs of context-sensitive reachability analysis with variable separation as presented in Section 3. Some configurations resulted in too strong over-approximations and therefore they could not prove safety of the given benchmarks (†), some others timed out (TO).

The benchmark instances fisher (4 × 1), platoon (2 × 1, 1 × 10), rods (3 × 1) and 2tanks (10 × 1, 6 × 2) can be decomposed into sub-spaces as indicated in the brackets (number of sets $V_i \times |Vi|$) while bball and sw5 are not decomposable. The mixed-rectangular toy example (toy) can be decomposed into 1 × 2, 1 × 1, 1 × 1 subspaces where the first subspace is rectangular, the second one is timed and the third one is linear. The rectangular version of fisher is decomposable into subspaces of dimension 1 × 3 and 1 × 1 where the first subspace is rectangular. Note that benchmarks involving rectangular subspaces together with non-rectangular subspaces always require decomposition, as the analysis method for rectangular automata fundamentally differs from the one for LHA II, which was normally used as a fallback in case no decomposition is demanded. Furthermore the state set representation for rectangular subspaces is fixed, as we use conjunctions of linear constraints to represent state sets. As the reachability analysis method for rectangular automata does not require time discretization but computes the set of reachable states in a location in one step, we do not consider any time step size for those benchmarks.

We can observe that for *support functions* the decomposition pays off in most cases in terms of running time, and the overhead introduced by the decomposition is negligible in comparison to the speed-up resulting from lower-dimensional sub-spaces. Even a state space reduction by two

Table 3: Running times (in seconds) for the rectangular models and adaptions. Note that per default decomposition is used.

Benchmark	Running Times
toy	0.11
fisher	20
5variable_system	19.4

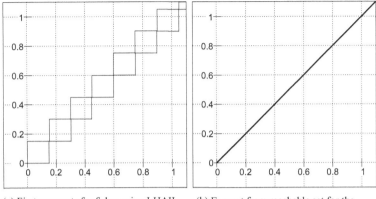

(a) First segments for fisher using LHAII reachability analysis methods ($\delta = 0.1$).

(b) Excerpt from reachable set for the rectangular variant of fisher.

Fig. 6: Plots of the computed set of reachable states for the first location in fisher using different analysis methods.

dimensions (platoon) is noticeable, the decomposition of fisher into one-dimensional sub-spaces allows to obtain results while the analysis using the original 4-dimensional state space exceeds the time limit. Note that for fisher the separated sub-spaces all require the usage of the linear context while the separated one-dimensional sub-spaces in platoon can both be computed using a timed context. The sub-spaces in 2tanks are mostly of discrete nature, which explains the huge speed-up when using support functions.

In our benchmarks, *boxes* behave differently. In general, boxes are amongst the fastest state set representations available such that the overhead introduced by decomposition as well as the overhead caused by instantiating multiple handlers computing flowpipes instead of one single handler is noticeable.

The analysis in the rectangular version of fisher is slower than when using boxes, however the state sets can be computed exactly with this method, which is not possible when using LHA II analysis methods (see Figure 6). The model of 5variable_system could be verified for one jump, when increasing the jump limit the memory limit was exceeded. This can be explained by the repeated application of Fourier-Motzkin variable elimination, which in the worst case introduces quadratically many new constraints when eliminating a variable.

5. Conclusion

We presented an extension to our previous work which is able to dynamically establish variable set decompositions based on syntactical independence. Furthermore our approach identifies a context for each sub-space which allows for applying specialized methods for reachability analysis via flowpipe construction, i.e., for timed automata or rectangular automata.

Future Work

The currently computed decompositions are computed for the whole automaton. As a further task it would be interesting to investigate on local decompositions, i.e., decompositions which are based solely on the current location, which allows for more dynamic approaches at the cost of additional over-approximation when switching between different decomposition schemes. In this work we created decompositions for rectangular subspaces solely based on syntactic features. For future work it would be interesting to provide an approach which allows to convert subspaces with constant derivatives to a rectangular subspace, provided guard and invariant conditions as well as resets comply with the properties of a rectangular automaton (conditions are axis-aligned, variables are only reset to intervals). Additionally using classic variable elimination techniques such as Fourier-Motzkin variable elimination may introduce quadratically many constraints among some are redundant. One direction for the future development would be to further develop automated reduction techniques or switch to more sophisticated approaches as for instance proposed in [11].

References

[1] Matthias Althoff and John M. Dolan. 2014. Online verification of automated road vehicles using reachability analysis. IEEE Transaction on Robotics 30(4): 903–918.

[2] Christel Baier and Joost-Pieter Katoen. 2008. Principles of Model Checking. The MIT Press.

[3] Stanley Bak and Marco Caccamo. 2013. Computing reachability for nonlinear systems with hycreate. Poster Session of HSCC'13.

[4] Stanley Bak and Parasara Sridhar Duggirala. 2017. Hylaa: A tool for computing simulation-equivalent reachability for linear systems. pp. 173–178. *In*: Proc. of HSCC'17, ACM.

[5] Johan Bengtsson, Kim Larsen, Fredrik Larsson, Paul Pettersson and Wang Yi. 1995. Uppaal—a tool suite for automatic verification of real-time systems. pp. 232–243. *In*: Proc. of HS'95, Vol.1066 of LNCS, Springer.

[6] Johan Bengtsson and Wang Yi. 2004. Timed automata: Semantics, algorithms and tools. pp. 87–124. *In*: Lectures on Concurrency and Petri Nets: Advances in Petri Nets, Springer, Berlin, Heidelberg.

[7] Sergiy Bogomolov, Marcelo Forets, Goran Frehse, Frédéric Viry, Andreas Podelski and Christian Schilling. 2018. Reach set approximation through decomposition with low-dimensional sets and high-dimensional matrices. pp. 41–50. *In*: Proc. of HSCC'18, ACM.

[8] Olivier Bouissou, Alexandre Chapoutot and Samuel Mimram. 2013. Computing flowpipe of nonlinear hybrid systems with numerical methods. CoRR, abs/1306.2305.

[9]   Marius Bozga, Conrado Daws, Oded Maler, Alfredo Olivero, Stavros Tripakis and Sergio Yovine. 1998. Kronos: A model-checking tool for real-time systems. pp. 546–550. *In*: Proc. of CAV'98, Vol. 1427 of LNCS.

[10]   Víctor A. Braberman, Alfredo Olivero and Fernando Schapachnik. 2002. ZEUS: A distributed timed model-checker based on KRONOS. pp. 503–522. *In*: Proc. of PDMC'02, Vol. 68:4 of Electronic Notes in Theoretical Computer Science, Elsevier.

[11]   Xin Chen, Erika Ábrahám and Goran Frehse. 2011. Efficient bounded reachability computation for rectangular automata. pp. 139–152. *In*: Proc. of RP'11, Vol. 6945 of LNCS, Springer.

[12]   Xin Chen, Erika Ábrahám and Sriram Sankaranarayanan. 2012. Taylor model flowpipe construction for non-linear hybrid systems. pp. 183–192. *In*: Proc. of RTSS'12, IEEE Computer Society Press.

[13]   Xin Chen, Erika Ábrahám and Sriram Sankaranarayanan. 2013. Flow*: An analyzer for non-linear hybrid systems. pp. 258–263. *In*: Proc. of CAV'13, Vol. 8044 of LNCS, Springer.

[14]   Xin Chen and Sriram Sankaranarayanan. 2016. Decomposed reachability analysis for nonlinear systems. pp. 13–24. *In*: Proc. of RTSS'16, IEEE Computer Society Press.

[15]   Pieter Collins, Davide Bresolin, Luca Geretti and Tiziano Villa. 2012. Computing the evolution of hybrid systems using rigorous function calculus. pp. 284–290. *In*: Proc. of ADHS'12, IFAC-PapersOnLine.

[16]   David L. Dill. 1990. Timing assumptions and verification of finite-state concurrent systems. pp. 197–212. *In*: Proc. of CAV'89, Vol. 407 of LNCS, Springer.

[17]   ParasaraSridhar Duggirala, Sayan Mitra, Mahesh Viswanathan and Matthew Potok. 2015. C2E2: A verification tool for stateflow models. pp. 68–82. *In*: Proc. of TACAS'15, Vol. 9035 of LNCS, Springer.

[18]   Andreas Eggers. 2014. Direct Handling of Ordinary Differential Equations in Constraint-solving-based Analysis of Hybrid Systems. Ph.D. thesis, Universität Oldenburg, Germany.

[19]   Martin Fränzle, Christian Herde, Stefan Ratschan, Tobias Schubert and Tino Teige. 2007. Efficient solving of large non-linear arithmetic constraint systems with complex Boolean structure. Journal on Satisfiability, Boolean Modeling and Computation 1: 209–236.

[20]   Frehse, G. and R. Ray. 2009. Design principles for an extendable verification tool for hybrid systems. pp. 244–249. *In*: Proc. of ADHS'09, IFAC-PapersOnLine.

[21]   Goran Frehse, Colas Le Guernic, Alexandre Donzé, Rajarshi Ray, Olivier Lebeltel, Rodolfo Ripado, Antoine Girard, Thao Dang and Oded Maler. 2011. SpaceEx: Scalable verification of hybrid systems. pp. 379–395. *In*: Proc. of CAV'11, Vol. 6806 of LNCS, Springer.

[22]   Willem Hagemann, Eike Möhlmann and Astrid Rakow. 2014. Verifying a PI controller using SoapBox and Stabhyli: Experiences on establishing properties for a steering controller. pp. 115–125. *In*: Proc. of ARCH'14, Vol. 34 of EPiC Series in Computer Science, EasyChair.

[23]   Thomas A. Henzinger. 1996. The theory of hybrid automata. pp. 278–292. *In*: Proc. of LICS'96, IEEE Computer Society Press.

[24]   Thomas A. Henzinger, Peter W. Kopke, Anuj Puri and Pravin Varaiya. 1998. What's decidable about hybrid automata? Journal of Computer and System Sciences 57(1): 94–124.

[25]   Kong, S., S. Gao, W. Chen and E.M. Clarke. 2015. dReach: δ-reachability analysis for hybrid systems. pp. 200–205. *In*: Proc. of TACAS'15, Vol. 9035 of LNCS, Springer.

[26]   André Platzer and Jan-David Quesel. 2008. KeYmaera: A hybrid theorem prover for hybrid systems (system description). pp. 171–178. *In*: Proc. of IJCAR'08, Vol. 5195 of LNCS, Springer.

[27] Stefan Ratschan and Zhikun She. 2005. Safety verification of hybrid systems by constraint propagation based abstraction refinement. pp. 573–589. *In*: Proc. of HSCC'05, Vol. 3414 of LNCS, Springer.

[28] Stefan Schupp and Erika Ábrahám. 2018. Efficient dynamic error reduction for hybrid systems reachability analysis. pp. 287–302. *In*: Proc. of TACAS'18, Vol. 10806 of LNCS, Springer.

[29] Stefan Schupp and Erika Ábrahám. 2018. Spread the work: Multi-threaded safety analysis for hybrid systems. *In*: Proc. of SEFM'18, Vol. 10886 of LNCS, Springer. To appear.

[30] Stefan Schupp, Erika Ábrahám, Ibtissem Ben Makhlouf and Stefan Kowalewski. 2017. HyPro: A C++ library for state set representations for hybrid systems reachability analysis. pp. 288–294. *In*: Proc. of NFM'17, Vol. 10227 of LNCS, Springer.

[31] Stefan Schupp, Johanna Nellen and Erika Ábrahám. 2017. Divide and conquer: Variable set separation in hybrid systems reachability analysis. pp. 1–14. *In*: Proc. of QAPL'17, Vol. 250 of EPTCS, Open Publishing Association.

[32] Stefan Schupp, Justin Winkens and Erika Ábrahám. 2018. Context-dependent reachability analysis for hybrid systems. pp. 518–525. *In*: Proc. of FMI'18, IEEE Computer Society Press.

[33] Walid Taha, Adam Duracz, Yingfu Zeng, Kevin Atkinson, Ferenc A. Bartha, Paul Brauner, Jan Duracz, Fei Xu, Robert Cartwright, Michal Konecný, Eugenio Moggi, Jawad Masood, Pererik Andreasson, Jun Inoue, Anita Sant'Anna, Roland Philippsen, Alexandre Chapoutot, Marcia O'Malley, Aaron Ames, Verónica Gaspes, Lise Hvatum, Shyam Mehta, Henrik Eriksson and Christian Grante. 2015. Acumen: An open-source testbed for cyber-physical systems research. pp. 118–130. *In*: Proc. of IoT 360° 2015.

Chapter **8**

Netflow Feature Evaluation for the Detection of Slow Read HTTP Attacks

Cliff Kemp, Chad Calvert and *Taghi M Khoshgoftaar**

1. Introduction

Network cyber attacks have become commonplace in today's world. These attacks have become very sophisticated and difficult to prevent. Many of the stealthy attacks target the application layer where they take advantage of vulnerabilities on web servers. Because web servers are open to the public they are accessed frequently by many users. The goal of attackers is to simulate legitimate, normal traffic as close as possible, which they do quite well. The task for those defending the networks is to determine the difference between normal and attack traffic. To make it even more of a challenge the attackers are always updating their attack methods.

One approach to assist the defenders of networks is machine learning. Networks have enormous amounts of data they collect. The data comes from various sources such as logs, full packet captures (FPCs), and Netflow traffic. Machine learning can use these sources with numerous machine learning algorithms. Also, algorithms have many options that can be used to optimize that algorithm for a given scenario. Additionally, there are techniques used to enhance the data before the machine learning algorithm is applied. Collecting

College of Engineering & Computer Science, Florida Atlantic University, Boca Raton, Florida.
Emails: ckemp1@fau.edu; ccalver3@fau.edu
* Corresponding author: khoshgof@fau.edu

the data is easy, however, the most difficult part of the machine learning process is selecting the most relevant attributes, commonly referred to as features, that will improve the performance of the machine learner. We must also keep in mind that a set of features that performs well with one machine learner may not work well with another. Discovering the correct set of features for machine learning is referred to as feature selection. The goal of feature selection is to determine the set of features that will produce the best accuracy and predictability for the machine learner. There are many types of network attacks, and in this paper we focus specifically on application layer Denial of Service (DoS).

DoS is an attack that aims to prevent normal communication with a resource by disabling the resource itself or an infrastructure device providing connectivity to it. DoS attacks have evolved and adapted to create a severe security threat to networks. Akamai's in-depth report, with insight into the latest Distributed Denial of Service (DDoS) and web application attacks, states that after two consecutive quarters of decline in total attacks, the number of DDoS attacks increased markedly in the second quarter of 2017 [6].

Recent years have brought a rise in application layer DDoS attacks targeting applications. They target not only the well-known Hypertext Transfer Protocol (HTTP) but also HTTPS, DNS, SMTP, FTP, VOIP, and other application protocols that possess exploitable weaknesses allowing for DDoS attacks. Much like attacks targeting network resources, attacks targeting application resources come in a variety of types including HTTP GET, Slow POST, and Slow Read. Slow Read approaches are particularly prominent, mostly targeting weaknesses in the HTTP protocol which, as the most widely used application protocol on the Internet, is an attractive target for attackers. Network resources are expected to provide seamless availability to employees for their day-to-day activities and to customers that purchase items and need access to online accounts twenty-four hours a day. Dependency for this access in networks today has become commonplace and as such has attracted malicious attackers who target these network servers, especially the web servers. The primary goal of DDoS attacks is to deny service provided to customers and employees.

There are distinct types of DDoS attacks, each performing at the various levels of the Open Systems Interconnection (OSI) model [20] as seen in Figure 1. The OSI model helps vendors create inter-operable network devices and software in the form of protocols so that different vendor networks could work with each other. The central concept of the OSI model is that the process of communication between two endpoints in a network divides into seven separate groups of related functions or layers. The application layer supplies network services to user applications. Network services are protocols that work with user data. For example, a web browser application uses the

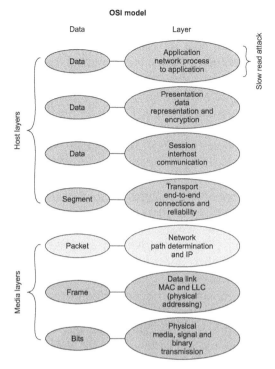

Fig. 1: OSI model.

application layer protocol HTTP which packages the data needed to send and receive web page content.

The Slow Read HTTP attack, also known as a "low and slow" attack [4], sends a legitimate HTTP request and reads the response slowly, aiming to keep as many connections active as possible to tie up resources on the server until it cannot handle any further requests. The characteristics of the Slow Read attacks relate to application resources, whereas the previous DDoS attacks targeted server resources such as bandwidth [50]. Slow Read attacks target specific application vulnerabilities, allowing an attacker to use a stealthy DoS. Not volumetric in nature, such attacks can often launch with only a single machine. Additionally, because these attacks occur at the application layer, a Transmission Control Protocol (TCP) handshake is already established, successfully making the malicious traffic look like normal traffic traveling over a legitimate connection.

Various evasion techniques are used to bypass intrusion detection systems, leaving the network vulnerable to specific DoS attacks. A stealthy DoS attack variant can disrupt routine web services covertly without triggering any alerts. One potential solution is the application of flow-based analysis.

Netflow, also referred to as session data, represents a high-level summary of network conversations. A Network flow record is identified based upon the standard 5-tuple attribute set that makes up a conversation: source IP, destination IP, source port, destination port, and transport protocol [32]. System for Internet-Level Knowledge (SiLK) is a software tool suite used to generate and analyze Netflow session data [19]. SiLK is a collection of traffic analysis tools developed by the Computer Emergency Response Team (CERT) and the Network Situational Awareness Team (NetSA) to facilitate security analysis of large networks [5]. SiLk can extract various standards of session data such as IP Flow Information Export (IPFIX) [13], Netflow v9 [12], or Netflow v5. It has the ability to collect Netflow session data in real-time or convert previously captured full pcaps. Netflow is a more space-efficient format than FPCs not so much because of its size, but because it records the packed records into service-specific binary flat files and can parse flows in a timely and efficient way without the need for complicated CPU intensive scripts [31]. This is a key factor when considering RAM and hard drive requirements for servers.

The first of three unique contributions of our experiment is the analysis of Slow Read HTTP DDos attacks and the use of machine learning predictive models that help detect Slow Read. An important aspect of our first contribution is the use of Netflow using the IPFIX standard for session data. The combination of Netflow session data (IPFIX) and machine learning combats these stealthy attacks by successfully responding to the evasive methods used by attackers. We take advantage of user-defined data types in its features, because of the IPFIX protocol being freely extensible and adaptable to different scenarios. Machine learning is applied to Netflow features with the following eight machine learning algorithms: Random Forest (RF), two variants of C4.5, a decision tree algorithm, 5-Nearest Neighbors (5NN), Multilayer Perceptron (MLP), JRip, which uses Repeated Incremental Pruning to Produce Error Reduction (RIPPER), Support Vector Machine (SVM), and Naïve Bayes (NB). We have chosen these machine learners based on their popularity with network traffic and the variations they represent [48]. These eight learners provide us with a comprehensive array of algorithms to use on our Netflow features.

Our second contribution focuses on the integrity of our data. Other studies have used computer-generated simulations, isolated test beds, or scripted traffic to collect their data [9], [39]. Our case study data is collected from real-world network data from a production computer network. Normal web server traffic was generated through interactions with students, faculty, and the public on our web server located on a college campus. This helps to produce results more representative of a real-world scenario. All attack data was produced on a live and currently active platform. The Slow Read attacks were generated by adjusting variables in the attack using three different levels of concurrent connections to give us a broad scope of this type of attack with represented models that reflect real-world activity. Therefore, the quality and integrity of our data is well represented because of the live setting instead of simulated environments or test beds. There are a few challenges when collecting data on a live network. These challenges are the generation of enough normal data, the concern for privacy, the generation of attacks on a live network for fear of disrupting the network, the number of machines needed to attack simultaneously, and the maintenance of network administration. Because of these challenges, other related works [45], [36] often utilize publicly available datasets.

Feature selection is our third contribution. We employ selective feature evaluation and investigate several methods used to specify the attribute evaluators and search methods. We evaluate the worth of a subset of attributes by considering the individual predictive ability of each feature along with the degree of redundancy between them. Subsets of features that are highly correlated with the class while having low intercorrelation are preferred. For this experiment, we choose the Weka [19] functions CfsSubsetEval and Consistency-SubsetEval. For single-attribute evaluation, we also used Weka functions ChiSquaredAttributeEval, GainRatioAttributeEval, and Principal Component Analysis (PCA). ChiSquaredAttributeEval and GainRatioAttributeEval are used with the Ranker search method to generate a ranked list from which Ranker discards a given number and ranks individual attributes according to their evaluation. Unlike other single-attribute evaluators, PCA transforms the given set of attributes into newly created subsets of its own.

The remainder of this paper is organized as follows. In Section 2, we detail a common Slow Read attack method and Netflow. In Section 3, we discuss related works associated with the collection and detection of a Slow Read attack and feature selection. Section 4 outlines our collection procedure, classification algorithms, and feature selection. In Section 5, we discuss our findings for feature selection and our learners. Lastly, in Section 6, we conclude our works and identify future endeavors.

2. Background

There are a variety of methods for enacting an application layer DDoS attack. Contingent on the characteristics of the network, various types of attacks are chosen based on the targeted traffic. Our experiment deploys a total of three different Slow Read application layer attacks with varying configurations to represent several levels of an attack. In this section, we detail the Slow Read attack, data collection process, and Netflow traffic.

In HTTP GET flood attacks, attackers can send different HTTP requests to the web server. The web server can have multiple connections from the same client to the same server. Each client process will be assigned a different ephemeral port number, so even if they all try to access the same server process, they will all have a different client socket and represent unique connections. This is what allows for several simultaneous requests to the same website from one computer. Attackers can target their requests toward the main web page, a random web page, a resource such as an image file, or even a combination of these [32]. Unlike high-bandwidth massive flooding attacks [4], low-bandwidth attacks performed by malicious users at the application layer rely on Slow Read attacks to evade detection. There is no need for an army of bots, as this type of attack can be performed with as little as one machine and use minimum bandwidth as compared to traditional flooding attacks [52]. Traffic during these attacks seems to be legitimate, where the HTTP client is a web browser that establishes a connection to a server for sending one or more HTTP request messages. We use an Apache web server as our HTTP server that accepts connections to serve HTTP requests by sending HTTP response messages. Differentiation of attack traffic and normal traffic is challenging and requires expertise in the field.

Application DDoS attacks most commonly target the HTTP protocol in an attempt to exhaust web servers through HTTP POST or GET requests. Dealing with DDoS flood attacks has simply been a matter of looking at overall flow volume for all routers to see if a spike had occurred. Once that determination has been made, administrators use methods to find the problem router or server and take steps to eliminate the threat. Attackers are increasingly targeting HTTP, DNS, and VoIP services to perform their attacks. Application DDoS attacks can target many different applications; however, the most common target which HTTP attacks aim to exhaust are web servers and services. Some of these attacks are characteristically more effective than others because they require fewer network connections to achieve their goal. For instance, an attacker could launch numerous HTTP GETs or POSTs to exhaust a web server or web application.

Slow Read attacks represent a method in which the attacker keeps the connection open by receiving the response from the server slowly, using a minimal TCP window size. TCP is the primary protocol of most modern networks, including the Internet. TCP is a reliable protocol that determines whether or not packets have been received and provides an ordered, and error-checked delivery of a stream of bytes between applications running on hosts. Part of the TCP specification RFC 1122 [3] allows a receiver to advertise a zero-byte window, instructing the sender to maintain the connection, but not send additional TCP payload data. The sender should then probe the receiver to check if the receiver is ready to accept data. By advertising a zero receive window and acknowledging probes, a malicious receiver can cause a sender to consume resources (TCP state, buffers, and application memory), preventing the targeted service or system from handling legitimate connections.

Figure 2 [1] shows the SlowHTTPTest tool performing a Slow Read attack which eventually purges a service's availability. The SlowHTTPTest tool efficiently implements various application layer DoS attacks. It executes most common low-bandwidth application layer DoS attacks, such as Slowloris, Slow HTTP POST, and Slow Read attack (based on TCP persist timer exploit) by draining the concurrent connections pool, as well as the Apache Range Header attack by causing significant memory and CPU usage on the server. If a data flow occurs, a Slow Read attack exploits the fact that most of the modern web servers are not limiting the connection duration. With the possibility of prolonging a TCP connection virtually forever with zero or minimal data flow by manipulating the TCP receive window size value, it is possible to acquire concurrent connection pools of the application [1]. It is the concurrent connections that will eventually bring the server down.

Unlike Slowloris and Slow HTTP Post in which attacks are performed by pushing data slowly to the server, a Slow Read attack forces the server to

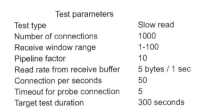

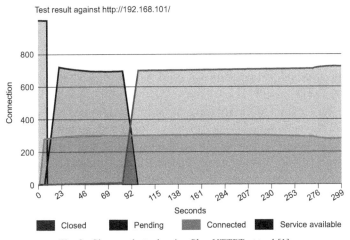

Fig. 2: Slow read attack using SlowHTTPTest tool [1].

send a significant amount of data which it accepts at a slow rate. The window size can be set to beyond 0 by the attacker arbitrarily. The attacker declares a very small receive window size which makes the server split the response into many small pieces that would fit the buffer size, resulting in prolonged ongoing responses [1].

This attack needs to have the target generate at least one piece of content on the web page that is larger than the buffer of the server. The attacker may load the main page of the website and pick the largest resource. If there are no sufficiently large resources on the server, the attacker will multiply the size of the response by repeatedly requesting the same resource which will fill up the send buffer of the web servers. The server will finish processing the request when the content has been stored in the send buffer, so if the content is smaller than the send buffer, the attack will fail.

Data collection represents a major contribution of our research, primarily due to the fact that the method of collection can have a direct impact on the quality of analysis and ability to perform effective attack detection. Network data comes in many formats and locations. Knowing what data is available, where that data comes from, how and why it is collected, and what can be done with it is a major responsibility for those who manage networks. FPCs, web server logs, and session data are all excellent sources for analyzing network traffic. Each of these data sources performs better than the other depending on the type of attacks one is examining. FPC data provides a full account for every data packet transmitted between two endpoints, but can be overwhelming due to its high degree of detailed data that is processed for diagnostics. Storing FPCs can be a challenge to keep for very long, if at all. Ideally, one has FPCs available for a shorter duration in case one must investigate any previous malicious activity. It comes at a cost, however, as it can be quite storage intensive to capture and store FPC data for an extended period. The large size prohibits most organizations from retaining any significant amount of data. Furthermore, evaluating all available packet features can be resource intensive. Some organizations do not have the resources to include FPC data into their Network Security Monitoring (NSM) infrastructure efficiently [44].

Web server logs are ineffective at early detection of Slow Read attacks because of the technique used by the Slow Read attack on a web server. Slow Read attacks keep TCP connections open with no data being sent. As long as the receiver TCP continues to send acknowledgments in response to the probe segments, the sender TCP must allow the connection to stay open. Given these conditions, the TCP connection will be open, with no data being transmitted. This "stalled" state is generally referred to as the TCP persist condition [22]. At this point, there are no web server logs generated during the attack. Web logs are only generated after the attack has completed, and the damage done.

As previously mentioned, SiLK can be used to supplement or supply a summary of some of the most valuable attributes of the traffic and maintain the data in a format that permits longer retention because of the significantly lower amount of data. Additionally, the features utilized in IPFIX Netflow are well designed to avoid evasion techniques used by attackers [44].

Unlike the challenges that FPC faces with accumulating and analyzing enormous amounts of data, Netflow session data simply includes a collection of text records and statistics. Session data is significantly smaller in size as compared to FPC. Session or flow records will usually include the protocol, source IP address and port, the destination IP address and port, a timestamp of when the communication began and ended, and the amount of data transferred between the two devices. Netflow represents an efficient storage solution for network data. This is beneficial for network security analysts as they must be able to rapidly query large historical traffic datasets. Additionally, Netflow is ideally suited for analyzing traffic on the backbone or border of a large, distributed enterprise, or mid-sized ISP.

The basic unit of data transfer in IPFIX is the message. A message contains a header and one or more sets, which contain records. A set may be either a template set, containing templates or a dataset, containing data records. A dataset references the template describing the data records within that set [21]. This is the mechanism which lends IPFIX its flexibility. IPFIX offers variable length fields for exporting custom information, where Netflow V9 does not. It also has a scheme for exporting lists of formatted data. Our Netflow data uses the IPFIX standard format on generated network traffic and, with the support of Silk NetFlow session data, performs comparably well. SiLK is a tool that can fill the gaps of traffic capture tools because it can find anomalies associated with traffic patterns and behaviors, and create statistics such as aggregate packets, duration, and bytes in a flow.

3. Related Works

Historically, network simulators like nse2 [27] or modelers such as Opnet [25] were used to reproduce DDoS attacks and measure their effects with attack detection techniques like [10] and [43]. Though these simulation methods were practical at that time, they are not a factual depiction of real-world environments [28]. Combined with slower speeds using traffic replay, this further illustrates why simulation is an ineffective scheme for focusing on DDoS attack detection techniques. A better solution to simulators is emulation. This is where actual machines are used as attackers and targets. One method of emulation is demonstrated with the Emulab DETER [9] and PlanetLab [39] testing environments. The DETER and Emulab testbeds permitted users to choose machines in a controlled facility that are inaccessible from the Internet. PlanetLab is a distributed test environment system that has shared access to machines, using Virtual Machine (VM) software, consequently isolating users within their own environment. Park et al. [38] analyze the effectiveness of Slow Read DoS attack using a virtual environment framework, but again their testbed is in an isolated setting.

As mentioned, we propose using the following machine learning techniques: RF, 5NN, MLP, C4.5N, C4.5D, JRip, SVMs, and NB. There have been other works that have explored some of these and other techniques. Adi et al. [7] used Weka to employ four machine learning techniques (NB, decision tree, JRip, and SVMs) and ranked features. Farnaaz et al. [16] used RF to conduct their experiment on the NSL-KDD dataset. Their results show that the proposed model is efficient with a low false alarm rate and high detection rate.

NB is one of the most widely used techniques in data mining communities and used in many studies on traffic analysis and DoS detection. Mukherjee et al. [30] evaluated datasets with NB, applying feature reduction using three standard feature selection methods: Correlation-based Feature Selection (CFS), Information Gain (IG), and Gain Ratio (GR). Another study [29] applied NB to classify traffic without inspecting the payload but rather by extracting features from the TCP headers. Najafabadi et al. [32] applied the PCA subspace anomaly detection method to analyze whether the proposed user behavior NB model can sufficiently distinguish between normal and attack instances.

Zhang et al. [53] proposed a technique to pre-process traffic before removing features to be classified using NB. The pre-processing technique correlated traffic flows that were generated by the same application. The study showed that the proposed method outperformed other machine learning techniques such as decision tree and K-Nearest Neighbors (KNN). Haddadi et al. [18] employed flow-based network traffic utilizing NetFlow (via Softflowd). The proposed botnet analysis system is implemented by employing two different machine learning algorithms, C4.5 and NB. Their study reported the use of decision trees to identify botnet behavior from generated traffic patterns. The scheme compared its performance analysis with NB and concluded that decision trees could produce better classification accuracies.

JRip is considered a faster machine learning technique than decision trees. Gaonjur et al. [17] used a JRip classifier in a traffic analysis experiment to reduce false alarms. To select the best traffic features Yang et al. [51]

used an algorithm to improve classification accuracy and reduce the cost of classification associated with using JRip. Panda et al. [37] developed an extended and repeated incremental pruning process method via JRip rule-based classifiers to construct multiple classifier systems to efficiently detect network intrusions.

SVMs have been applied to classify DoS traffic and legitimate traffic in a recent study [47]. Their study showed that SVMs classify with higher than 90% accuracy in all conducted experiments. Najafabadi et al. [33] apply the one-class SVM algorithm on the extracted features from normal users' HTTP request sequences and label any newly seen instance that deviates from the normal trained model as an application layer DDoS instance.

Feature selection has been proven to be effective and efficient in preparing data for various data mining and machine learning problems [23]. The objectives of feature selection include building simpler and more comprehensible models, improving data mining performance, and preparing clean, understandable data. PCA is a popular tool for data analysis and dimensionality reduction. A disadvantage is the fact that the principal components are usually linear combinations of all variables where all weights in the linear combination, are typically non-zero. To solve this disadvantage, [15] applies Sparse Principal Component Analysis (SPCA) to select features that can retain the total variance maximally by considering interactions among features and can select features with less redundancy.

Najafabadi et al. [34] used existing Netflow features with C4.5 decision tree algorithms. The decision tree algorithms build the predictive rules by using the dataset attributes. This provides a way to interpret how the newly defined attributes are contributing in the detection of attacks using their four models that include two versions of C4.5 decision tree, NB, and 5NN algorithms.

One benefit of feature selection is the decreasing of process time. [35] selected four filter-based feature selection methods which are chosen from two categories for the application of network intrusion detection. Their methods, which consist of three filter-based feature rankers and one filter-based subset evaluation technique, are compared together along with the null case which applies no feature selection. They apply statistical analysis to determine whether performance differences between these feature selection methods are significant or not. Bauder et al. [8] present two case studies with medical claims fraud by employing a RF model with random undersampling, to mitigate adverse effects of class imbalance and to generate seven different class distributions for several big Medicare claims datasets. [46] provides a unique insight into the underlying relationships among classifier performance metrics by applying factor analysis to the classifier performance space. This provides an improved understanding about relationships and groupings with performance metrics, facilitating the selection of performance with those metrics.

4. Experimental Procedure

The outline of our experiment concerning Slow Read attacks in this section is divided into three subsections that include the data collection process, classification algorithms, and metrics used to evaluate each model.

4.1 Data Collection Process

Our capture framework allows for us to perform our attacks within a real-world network environment servicing numerous active users. The campus network consists of hosts from classrooms, labs (including off-campus, virtual systems), and offices. To facilitate our network usages, we employ switches, servers, and routers capable of servicing on-and-off-campus users. A Cisco firewall is used to provide secure access to data and network resources. For a student resource portal, an Apache web server has been set up to serve as the target for our attacks. The configuration of the server is composed of a Linux CentOS operating system, an Intel 3.30 GHz processor, and 32 GB of memory. Figure 3 shows our architecture in more detail.

We installed WordPress on the Apache student web server to serve as our content management system. The website consists of lecture material, assignments, assessments, and other content required by student users. Normal traffic related to coursework may consist of but is not limited to:

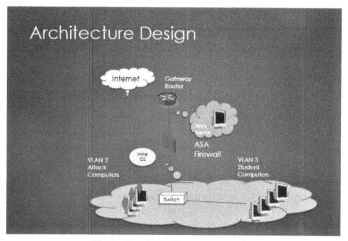

Fig. 3: Topology of campus network.

downloads, uploads, website navigation, and other communications with the web server. Within the context of our network usage, students both locally on our network and from online may request course material concurrently from our server. Our extended network also supports other faculty and students by providing services such as virtualization, email, web hosting, and audio/video streaming. Additional traffic is generated from the public as this is a live web server facing the Internet.

Our attacks are performed using penetration testing on a physical host machine using the SlowHTTPTest [2] tool rather than through simulation. Variations in settings used by SlowHTTPTest are applied, giving us different results and valuable information on thresholds of the attacks in our experiments. For our tests, the SlowHTTPTest tool allowed for easy configuration adjustments and incorporated numerous attack settings. We administered a total of three different attacks with varying configurations to represent several levels of an attack. For an attack attempting to avoid detection by using minimal connections, representing a stealthy scenario, we configured a single attack host using 500 connections with a random connection interval between 1 and 5 seconds. Our second attack initiated a more moderate level attack executing 1,000 connections with the same random connection interval between 1 and 5 seconds. Finally, we implemented the least stealthy attack, using 1,500 connections and again the random connection interval was set between 1 and 5 seconds. Each attack ran from a single host machine for approximately one hour and targeted our resource web server.

As previously mentioned, FPC can be quite storage and resource intensive in capturing FPC data for an extended period of time. Calvert et al. [11] collect data using web server logs from a student resource web server comprised of 29 unique fields. Web server logs are ineffective at early detection of Slow Read attacks because of the techniques used by the Slow Read attack on our web server. Netflow data is just a collection of text records and statistics and is incredibly small in size. The smaller size allows faster parsing and analyzing of data. The result is that it is easy to create large-scale flow storage solutions. FPC data retention is in minutes or hours, but Netflow data can be retained for months or years. Analysis tools aid analysts in examining the data for the purpose of detecting anomalies or generating statistics.

There are a few Netflow standards used on networks today. We use the IPFIX standard for our Netflow data. As mentioned previously, the IPFIX standard is chosen because of its flexibility and features. It offers variable length fields for exporting custom information, where NetFlow V9 does not and can take advantage of user-defined data types in its messages, so the protocol is freely extensible and can adapt to different scenarios. The features used in our work for machine learning are listed in Table 1.

Table 1: Description of netflow features.

Feature Name	Description
Protocol	IP protocol
Packets	Number of packets in flow
Bytes	Number of bytes in flow
Flags	TCP flags all packets [FSRPAUEC]
InitialFlags	TCP flags in initial packet
SessionFlags	TCP flags second through final packet
Attributes	Flow attributes [SFTC]
Packets/Second	Number of packets per second
Bytes/Second	Number of bytes per second
Bytes/Packet	Number of bytes per packet
Durmsec	Duration of the flow (in seconds)
Label	Class label (Attack or Normal)

4.2 Classification Algorithms

Eight classification algorithms were selected to build predictive models based on our collected datasets: RF, two variants of C4.5 decision trees (C4.5N, C4.5D), 5NN, MLP, JRip, SVM, and NB. This variety of learners was selected to broaden the scope of our analysis. All models were built using the Weka machine learning toolkit [19]. Weka contains implementations of machine learning algorithms used in this research [49]. The Weka Java API is used to write the framework. The software versions used are 3.6:14 for Weka and Java JDK version 8u91.

Machine learning algorithms can be grouped into parametric and nonparametric models. Using parametric models, we estimate parameters from the training dataset to learn a function that can classify new data points without requiring the original training dataset. Two examples of parametric models used in our experiment are the MLP and SVM. In contrast, nonparametric models cannot be characterized by a fixed set of parameters, and the number of parameters grows with the training data. Four of our nonparametric models used in our work are the decision tree classifiers C4.5D, C4.5N, 5NN, and RF.

5NN is a specific value for KNN, a typical example of a lazy learner, which means it does not learn a discriminative function from the training data, but memorizes the training dataset instead. KNN is described as instance-based learning that performs predictions by finding the prediction value of records (near neighbors). These distance functions utilize K, which represents the number of closest instances to the test instance to decide its label. The KNN algorithm itself is straightforward and can be summarized by the following steps:

1. Choose the number of K and a distance metric.
2. Find the K-nearest neighbors of the sample that we want to classify.
3. Assign the class label by majority vote.

Figure 4 illustrates how a new data point represented by the circled question mark is assigned the triangle class label based on majority voting among its five nearest neighbors.

The C4.5 decision tree is a tree-based learning algorithm which is used for classification problems. The C4.5 algorithm for building decision trees is implemented in Weka as a classifier called J48. The classifiers are organized in a hierarchy and model training is used to learn parameters from the data. C4.5 selects the attribute of the data that most efficiently splits its set of samples into subsets augmented in one class or the other. The splitting criterion is the normalized IG. The attribute with the maximum normalized IG is selected to make the decision. We utilized a version of C4.5 using the default parameter values from Weka (denoted by C4.5D) as well as a version (denoted by C4.5N) with Laplace smoothing activated and tree-pruning deactivated.

RF has gained huge popularity in applications of machine learning during the last several years due to its good classification performance, scalability, and ease of use. RF is an ensemble classifier used to improve the accuracy as compared to a single decision tree. Intuitively, RF can be considered as an

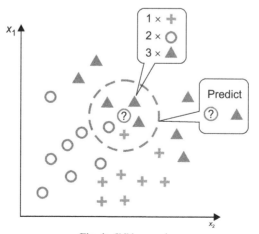

Fig. 4: 5NN example.

ensemble of decision trees. The idea behind RF is to average multiple (deep) decision trees that individually suffer from high variance. This will help to build a more robust model that has a better generalization performance and is less susceptible to overfitting. It corrects the tendency of decision trees to overfit by randomly sampling features as split candidates during each iteration. When constructing individual trees in RF, randomization is applied to select the best node to split on. In our experiment, no changes were made to the default values for RF. The RF algorithm can be summarized in four simple steps:

1) Draw a random bootstrap sample of size n (randomly choose n samples from the training set with replacement).
2) Grow a decision tree from the bootstrap sample. At each node:
   - a. Randomly select d features without replacement.
   - b. Split the node using the feature that provides the best split according to the objective function.
3) Repeat steps 1 and 2 K times.
4) Aggregate the prediction by each tree to assign the class label by majority vote.

Algorithm 1 RF Algorithm

1: **repeat** $t >$ RF
2:     **for** Draw random bootstrap sample of size n (randomly choose n samples from training set with replacement) **do**
3:         Grow a decision tree from the bootstrap sample. At each node:
4:         a. Randomly select d features without replacement
5:         b. Split the node using the feature that provides the best split according to the objective function
6:         Repeat steps 1 and 2 K times
7:         Perform majority vote across trees
8:     **end for**
9:     Aggregate the prediction by each tree to assign the class label

---

An MLP is a type of artificial neural network that utilizes neurons (perceptron) to compute an individual value from multiple inputs using nonlinear transformations. Although a single neuron can be somewhat limiting, MLPs use these neurons as building blocks to create much larger networks. MLPs can also utilize several hidden layers to help transform inputs into usable values by the outer layers. Two parameters were changed from the default values for the MLP learner. The "hiddenLayers" parameter was changed to "3" to define a network with one hidden layer containing three nodes. The "validationSetSize" parameter was changed to "10" causing the classifier to leave 10% of the training data aside. This parameter is used as a validation set to determine when to stop the iterative training process. The MLP algorithm is demonstrated below:

Algorithm 2 MLP Algorithm

 1: **repeat** $t >$ MLP
 2:    **for** each training vector pair (x,t) **do**
 3:        evaluate the output y when x is the input
 4:        **if** $y \neq t$ **then**
 5:           form a new weight vector w' according to
 6:           $w' = w + a\,(t{-}y)\,x$
 7:        **else**
 8:           do nothing
 9:        **end if**
10:    **end for**
11: **until** $y = t$ for all training vector pairs = 0

JRip is based on rule-learning techniques that classify data samples into a single class and seeks a set of rules to classify data best. JRip has four principle stages, namely initialization stage, building stage that involving growing and pruning steps, optimization stage and the deletion stage. Classes are examined in increasing size and an initial set of rules for the class is generated using incremental reduced error. JRip proceeds by treating all the examples of a judgment in the training data as a class and finding a set of rules that cover all the members of that class. Thereafter it proceeds to the next class and does the same, repeating this until all classes have been covered. The rule-learning technique is what makes it a faster machine learning technique than decision trees. JRip reduces false alarms, selects the best traffic features, and efficiently reduces the volume of data processed by intrusion detection systems for classification [14]. Note that p and n are the number of true and false positives respectively. P and N are the total number of real positives and real negatives respectively. T is the number of instances and t is the number of examples of a selected attribute. The algorithm is briefly described as follows:

Algorithm 3 JRip Algorithm

 1: Initialize RS = {}

 2: **for** each class from the less prevalent one to the more frequent one **do**

 3:    Building stage:

 4:    **repeat**

 5:       a. Grow phase: Grow one rule by greedily adding antecedents (or conditions) to the rule until the rule is 100 percent accurate. The procedure tries every possible value of each attribute and selects the condition with highest IG: $p(\log(p/t)–\log(P/T))$

 6:       b. Prune phase: Incrementally prune each rule and allow the pruning of any final sequences of the antecedents; the pruning metric is $(p–n)/(p+n)$—but it is actually $2p/(p+n) –1$, so in this implementation we simply use $p/(p+n)$ (actually $(p+1)/(p+n+2)$, thus if $p+n$ is 0, it is 0.5).

 7:    **until** the description length (DL) of the ruleset and examples is 64 bits greater than the smallest DL met so far, or there are no positive examples, or the error rate is greater than or equal to 50 percent.

 8:    Optimization stage: After generating the initial ruleset Ri, generate and prune two variants of each rule Ri from randomized data using procedure a and b. One variant is generated from an empty rule, while the other is generated by greedily adding antecedents to the original rule. Moreover, the pruning metric used here is $(TP+TN)/(P+N)$. Then the smallest possible DL for each variant and the original rule is computed. The variant with the minimal DL is selected as the final representative of Ri in the ruleset. After all the rules in Ri have been examined and if there are still residual positives, more rules are generated based on the residual positives using Building Stage again.

 9:    Delete the rules from the ruleset that would increase the DL of the whole ruleset if it were in it and add resultant ruleset to RS.

10: **end for**

SVMs are discriminate classifiers used as supervised learning models with associated learning algorithms which can be considered an extension of the perceptron. SVMs utilize hyperplanes to separate instances in a dataset into two distinct groups and assign new instances to one class or another. The aim is for the SVM to identify the most optimal hyperplane with the most significant gap between class instances as possible. The margin is defined as the distance (W^T) between the separating hyperplane (decision boundary) and the training samples (X) that are closest to this hyperplane, which are called support vectors. This is illustrated in Figure 5 [41].

Some of the advantages of SVM is that it works well with a clear margin of separation and is effective in high dimensional spaces. It is also useful in cases where some dimensions are greater than the number of samples. SVM is also memory efficient uses a subset of training points in the support vectors. A couple of disadvantages is that it does not perform well when there are large data sets because of the required training time, and there is more noise when target classes are overlapping. SVM is only directly applicable for two-class tasks.

NB is based on the Bayesian theorem and is well-suited for datasets with high dimensionality. The algorithm works upon the assumption that features are independent and utilizes this premise to calculate the posterior probability that an instance is a member of a specific class. Being relatively robust, easy to implement, fast, and accurate, NB classifiers are used in many different fields. The independence assumption is often violated, but NB classifiers still tend to perform very well under this unrealistic assumption [40]. For small sample sizes, NB classifiers can outperform the more powerful alternatives. In some cases, NB with feature reduction is known to outperform other sophisticated classification methods [30]. However, strong violations of the independence assumptions and non-linear classification problems can lead to very poor performances of NB classifiers when random samples have a lack of independence and relevance of the variables.

Figure 6 demonstrates why NB performs better with linear problems (A), as opposed to non-linear problems (B). Random samples for two different classes are shown as colored spheres and the dotted lines indicate the class boundaries that classifiers try to approximate by computing the decision boundaries. A non-linear problem (B) would be a case where linear classifiers, such as NB, would not be suitable since the classes are not linearly separable [40]. In such a scenario, non-linear classifiers like KNN would be preferred.

4.3 Feature Evaluation

Selective feature evaluation uses several methods to specify the attribute evaluator and search methods. Attribute selection is normally done by searching the space of attribute subsets, evaluating each one by combining 1 of the 6 attribute subset evaluators with 1 of the 10 search methods. Subset evaluators take a subset of attributes and return a numerical measure that guides the search. In our experiment we choose the Weka functions CfsSubsetEval and ConsistencySubsetEval. For single-attribute evaluation we also used Weka functions ChiSquaredAttributeEval, Gain-RatioAttributeEval and PCA. ChiSquaredAttributeEval and Gain-RatioAttributeEval are used with the

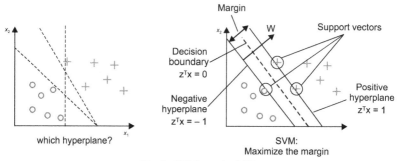

Fig. 5: SVM margins [41].

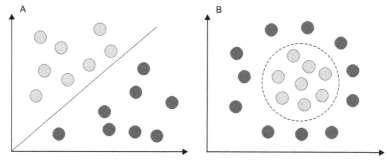

Fig. 6: Linear (A) vs. non-linear problems (B) for NB.

Ranker search method to generate a ranked list from which Ranker discards a given number. Unlike other single-attribute evaluators, PCA transforms the set of attributes.

1) Attribute Subset Evaluation

Correlation Feature Selection (CFS) evaluates the worth of a subset of attributes by considering the individual predictive ability of each feature along with the degree of redundancy between them. Subsets of features that are highly correlated with the class while having low inter-correlation are preferred. In Weka, CfsSubsetEval has an option that iteratively adds attributes that have the highest correlation with the class, provided that the set does not already contain an attribute whose correlation with the attribute in question is even higher. BestFirst and Greedy Stepwise are the search methods used in our work.

ConsistencySubsetEval evaluates attribute sets by the degree of consistency in class values when the training instances are projected onto the set [24]. Subsets of features that are highly correlated with the class while having low intercorrelation are preferred. Consistency of any subset can never be lower than that of the full set of attributes. Hence, the usual practice is to use this subset evaluator in conjunction with a Random or Exhaustive search which looks for the smallest subset with consistency equal to that of the full set of attributes [24]. The correlation of subsets is based on a merit between 0 and 1. The merit function will have larger values for attribute subsets that have attributes with strong class-attribute correlation and weak attribute-attribute correlation.

In our experiment, this evaluator is used in conjunction with BestFirst-Forward (BFF), BestFirst-Backward (BFB), Exhaustive Search (ES) and Random Search (RS) methods. BestFirst performs greedy hill climbing with backtracking; the number of consecutive non-improving nodes that must be encountered before the system backtracks can be specified. BFF can search forward from the empty set of attributes, backward from the full set, or start at an intermediate point (specified by a list of attribute indexes) and search in both directions by considering all possible single-attribute additions and deletions.

RandomSearch randomly searches the space of attribute subsets. If an initial set is supplied, it searches for subsets that improve on (or equal) the starting point and have fewer (or the same number of) attributes. Otherwise, it starts from a random point and reports the best subset found. The fraction of the search space to explore can be determined. ExhaustiveSearch searches through the space of attribute subsets, starting from the empty set, and reports the best subset found. If an initial set is supplied, it searches backward from this starting point and reports the smallest subset with a better (or equal) evaluation.

2) Single-Attribute Evaluation

Single-attribute evaluators are used with the Ranker search method to generate a ranked list from which Ranker discards a given number. Ranker is not a search method for attribute subsets, but a ranking scheme for individual attributes. It sorts attributes by their individual evaluations and must be used in conjunction with one of the single-attribute evaluators and performs attribute selection by removing the lower ranking ones.

In our work, we used Chi-Squared, GR, IG, and Symmetrical Uncertainty. Chi-squared value-based feature selection computes the Chi-squared statistical value for all features with respect to each class and ranks the features based on the value. The algorithm poses an initial hypothesis that a class and a feature are unrelated. Then, it works towards disproving the initial hypothesis.

GR is another information theory-based feature ranking technique where the IG score for a given feature is normalized by the Information Split value or Intrinsic Value of the feature. Information Split value or Intrinsic Value is the entropy measure of the attribute using various rank-based algorithms. IG is an information theory-based feature ranking technique that measures the extent of information possessed by a feature. The algorithm computes the scope of a feature towards the entropy of a class. Good features reduce the entropy of a class to the maximum level. Symmetric Uncertainty is another information theory-based feature ranking technique where the IG score is normalized by the entropy value of the attribute and the class. A good feature should have a high score.

3) Principal Component Evaluation

As mentioned previously, PCA transforms the set of attributes. New attributes are ranked in order of their eigenvalues. Technically speaking, the amount of variance retained by each principal component is measured by the so-called eigenvalue. A subset is selected by choosing enough eigenvectors to account for a given proportion of the default variance of 95 percent. The dimensionality of our dataset needs to be reduced by compressing it onto a new feature subspace, by selecting the subset of the eigenvectors, otherwise known as principal components, which contain most of the information that makes up the variance. The eigenvalues define the magnitude of the eigenvectors. We then sort the eigenvalues in decreasing order and focus on the top k eigenvectors based on of their corresponding eigenvalues.

PCA is a linear transformation technique that is widely used across different fields, most prominently for feature extraction and dimensionality reduction. PCA aims to find the directions of maximum variance in high-dimensional data and projects it onto a new subspace with equal or fewer dimensions than the original one. The orthogonal axes (principal components) of the new subspace can be interpreted as the directions of maximum variance given the constraint that the new feature axes are orthogonal to each other.

PCA also helps us to process our data for a technique called T-distributed Stochastic Neighbor Embedding (t-SNE) that visualizes high-dimensional data by giving each data point a location in a two or three-dimensional map [42]. The technique is a variation of Stochastic Neighbor Embedding that is much easier to optimize, and produces significantly better visualizations by reducing the tendency to crowd points together in the center of the map. [26] suggests that t-SNE is better than existing techniques at creating a single map that reveals structure at many different scales.

4.4 Metrics

We evaluate each model using the Area Under the receiver operating characteristic Curve (AUC) and Precision-Recall metrics. The AUC graphs the True Positive Rate (TPR) and False Positive Rate (FPR) of the model. TPR represents the percentage of the Slow Read attack instances correctly predicted as an Attack label. FPR represents the percentage of the normal data which was wrongly predicted as an Attack label. The AUC curve is built by plotting TPR vs. FPR as the classifier decision threshold is varied. Higher AUC values tend to correlate to higher TPR and lower FPR, both of which are preferred outcomes. Since this data was generated by network equipment, missing values were not present.

Stratified five-fold cross-validation with four iterations is used to evaluate our AUC values. Stratified five-fold cross-validation divides the data into five non-overlapping parts, with original class ratios being maintained in each fold. For each iteration, one part is kept as test data and the remaining four elements used as training data. Our final AUC values are calculated by aggregating the AUC values of the models tested for each of five elements of the data. Our experiment applied four runs of five-fold cross-validation to provide average performance values and decrease the bias of randomly selected folds.

There are two kinds of errors we use for metrics. A Type 1 error measures the total amount of false positives. A false positive (FP) is when the outcome is incorrectly predicted as yes (positive) when it is no (negative). A Type 2 error measures the total amount of false negatives. A false negative (FN) is when the outcome is incorrectly predicted as negative when it is positive. They are both important misclassification errors that should be minimized, but the emphasis should be more on the Type 2 error. The reason for this is that in a network it is more important not to miss an attack as opposed to identifying an attack that is not. If a Type 2 error occurs, an attack has not been identified and the network has been compromised whereas, mislabeling normal traffic as an attack is not as severe as a missed attack.

Precision-Recall is a useful measure of success of prediction when the classes are very imbalanced. In information retrieval, precision is a measure of result relevancy, while recall is a measure of how many truly relevant results are returned. The F-measure (F-score), which is a measure of a test's accuracy, is defined as the weighted harmonic mean of the precision and recall of the test and conveys the balance between the precision and the recall. An F-score reaches its best value at 1 (perfect precision and recall) and worst at 0. High scores show that the classifier is returning accurate results (high precision), as well as returning a majority of all positive results (high recall). A system with high recall but low precision returns many results, but most of its predicted labels are incorrect when compared to the training labels. A system with high precision but low recall is just the opposite, returning very few results, but most of its predicted labels are correct when compared to the training labels. An ideal system with high precision and high recall will return many results, with all results labeled correctly.

Analysis of variance (ANOVA) is a collection of statistical models and their associated procedures (such as "variation" among and between groups) are used to analyze the differences among group means. A one-factor ANOVA is used to compare means from two independent (unrelated) groups using the F-distribution. The null hypothesis for the test is that the two means are equal. Therefore, a significant result means that the two means are unequal. We used ANOVA to compare AUC means of a Slow Read attack detection among the eight learners and check if differences are statistically significant. Tukey's Honestly Significant Difference (HSD) post hoc test is used on our data in conjunction with ANOVA to find means that are significantly different from each other. The Tukey's HSD post hoc test compares all possible pairs of means to find out which specific group means (compared with each other) are different.

5. Results

The following sub-sections include our results for feature selection and machine learner performance. Our analysis presents a comprehensive comparison of the different feature selection methods for detecting Slow Read DDoS attacks. We compare four different feature selection methods and four different classifiers. Our feature selection methods and classifiers are used to build classification models. Results are also given for the original eleven-feature dataset. Our goal is to achieve the same performance or better than that of the eleven-feature set using feature selection methods. This is accomplished by focusing on the total number of Type 1 and Type 2 errors, with more weight assigned to the Type 2 errors. Precision, recall, and F-measure are utilized to compare performance of the models.

5.1 Feature Selection

CFS evaluators in Table 2 display the merit value, number of features selected, and the names of these selected features. The CFS evaluators are denoted as BestFirst-Forward (BFF), BestFirst-Backward (BFB), GreedyStepwise-Forward (GSF), GreedyStepwise-Backward (GSB), RankSearch-ChiSquared (RSCS), and RankSearch-GainRation (RSGR). The evaluation results on each CFS subset evaluator is provided in Table 3. The number on the end of each model represents number of features. The results indicate that all four classifiers (RF, C4.5N, C4.5D, 5NN) show good performance with the five-feature subset of Bytes, Flags, Initial Flags, Bytes/Packet, and Durmsec. The performance metrics of C4.5N were slightly better than C4.5D. Although C4.5N without pruning will construct a tree that is more profound and complex, which will produce a longer and more complicated tree structure. C4.5N had the same number of Type 2 errors as the eleven-feature set. RF and 5NN had slightly less Type 2 errors as compared to their original eleven-

Table 2: CFS subset evaluator methods.

Evaluators	Merit	Features	Features
BFF	0.183	3	Flags, Bytes/Packet, Durmsec
BFB	0.183	3	Flags, Bytes/Packet, Durmsec
GSF	0.183	3	Flags, Bytes/Packet, Durmsec
GSB	0.181	5	Bytes, Flags, InitialFlags, Bytes/Packet, Durmsec
RSCS	0.181	2	Bytes, Flags
RSGR	0.181	5	Bytes, Flags, InitialFlags, Bytes/Packet, Durmsec

Table 3: CFS model results.

Models	Type 1	Type 2	Prec.	Recall	F-Meas
RF3	9446	706	0.708	0.970	0.818
RF5	9444	700	0.708	0.970	0.818
RF11	9446	706	0.708	0.970	0.818
C4.5N3	9477	711	0.707	0.970	0.818
C4.5N5	9470	709	0.707	0.970	0.818
C4.5N11	9470	709	0.708	0.970	0.818
5NN3	9538	761	0.705	0.968	0.816
5NN5	9523	742	0.706	0.969	0.816
5NN11	9519	748	0.706	0.968	0.816
C4.5D3	9480	724	0.706	0.969	0.817
C4.5D5	9478	720	0.707	0.969	0.818
C4.5D11	9478	720	0.706	0.968	0.816

feature set. The RF five-feature set had the least amount of Type 2 errors overall. Precision, recall, and F-measure had similar results for all the models. The CFS Feature Selection technique provides effective performance with less features, by reducing the feature space to the most relevant features (also reducing the chance of overfitting and decreases processing time). The five-feature subset is effective for the detection of Slow Read attacks with all four classifiers.

The results for our second attribute subset evaluator, ConsistencySubset Eval, are illustrated in Tables 4 and 5. Table 4 displays the attribute selection results for two search methods (Random and Exhaustive Search) and two options (forward and backward search). They are denoted as, BestFirst-Forward (BFF), BestFirst-Backward (BFB), Exhaustive Search (ES), and Random Search (RS). The search algorithm selected particular features denoted by "X", and excluded a particular feature denoted by "0". Table 5 shows the results for our four clasifiers (RF, C4.5N, C4.5D, 5NN). Each classifier has

Table 4: Consistency subset evaluator feature selection–attributes selected and excluded.

Features	BFF	BFB	ES	RS	Total
Protocol	0	0	0	X	1
Packets	X	0	X	0	1
Bytes	0	X	0	X	2
Flags	X	0	X	X	3
InitFlags	0	X	0	0	1
SessFlags	0	X	0	0	1
Attributes	0	0	0	0	0
Packt/Sec	X	X	X	X	4
Bytes/Sec	X	X	X	X	4
Bytes/Packt	X	X	X	X	4
Durmsec	X	X	X	X	4
Total	6	7	6	7	0

Table 5: Consistency subset evaluator model results.

Models	Type 1	Type 2	Prec.	Recall	F-Meas
RF6	9475	734	0.707	0.969	0.817
RF7	9445	716	0.708	0.970	0.818
RF11	9446	706	0.708	0.970	0.818
C4.5N6	9495	746	0.706	0.968	0.817
C4.5N7	9479	716	0.707	0.970	0.818
C4.5N11	9470	709	0.707	0.970	0.818
5NN6	9558	766	0.705	0.968	0.816
5NN7	9558	766	0.705	0.968	0.815
5NN11	9519	748	0.706	0.968	0.816
C4.5D6	9498	753	0.706	0.968	0.817
C4.5D7	9481	724	0.707	0.969	0.817
C4.5D11	9478	720	0.707	0.969	0.818

three feature subsets. These are the set of eleven features (denoted by 11), the set using ConsistencySubsetEval with RandomSearch (denoted by 7), and the set with our top six overall scoring features using BFF and ES (denoted by 6). ConsistencySubsetEval with RandomSearch had the best results out of the four when we ran our group A classifiers (RF, C4.5N, 5NN, C4.5D). The customized seven-feature set in not in Table 5.

The results indicate that all the four classifiers (RF, C4.5N, C4.5D, 5NN) show good performance with the seven-feature subset of Protocol, Bytes, Flags, Packets/Sec, Bytes/Sec, Bytes/Packet, and Durmsec based upon error rates and Precision-recall. The seven-feature set had slightly less Type 2 errors as compared to the six-feature set, except for 5NN which was the same. We re-examined the tree structure of the original eleven and noticed that Packets and Bytes were discriminate factors. The seven feature set did not include Packets, but instead included Protocol. In almost all of our other feature selection methods, Protocol was not selected in the feature subsets. We took the feature Protocol and replaced it with Packets, causing C4.5D to obtain better results in all metrics. RF and C4.5N had the lowest number of Type 2 errors followed by C4.5D and 5NN which had the most. The revised seven-feature set maintained satisfactory precision, recall, and F-measure results, verifying the predictability of the seven features. The Consistency Feature selection is effective in producing less features that discriminate between normal and attack data for Slow Read attacks.

For the new seven-feature subset, RF, C4.5N, and C4.5D all performed slightly better than the original eleven features. With our feature adjustment to the Consistency subset, we produced a slightly better classification performance. Decision trees were helpful in identifying the Packets feature to replace Protocol feature, that helps to improve performance. This new seven-feature set produces results that improves the ability to discriminate between classes successfully and remove irrelevant and redundant features. These improvements help reduce the chance of overfitting and decrease process time.

In Table 6 we again took our top four classifiers and compared the five and seven-feature sets with the eleven-feature set. The five and seven-feature sets were produced from the four single-attribute evaluators; GR, Chi-Squared, Symmetric-Uncertainty, and IG.

RF had the lowest Type 1 and Type 2 error results with five features using the GR evaluator. The five-feature set consisted of Flags, Bytes, Bytes/Packet, Durmsec, and InitialFlags. Both RF and 5NN achieved slightly less Type 2 error results versus the full eleven-feature set. All subsets had good results with precision, recall, and F-measure metrics. All four classifiers provide good performance by using five and seven subset features. C4.5D results were not as favorable as C4.5N. As mentioned previously in this paper, the main difference between the default settings of C4.5D and C4.5N is that the latter is used with Laplace smoothing activated, and tree-pruning deactivated. The single-attribute method optimizes results using the GR attribute evaluator with five features for RF and 5NN and seven features for C4.5N and C4.5D. As mentioned previously, the benefits of reducing features from the original eleven-feature set include better processing times, less chance of overfitting, and a reduction of FNs during the detection of Slow Read DDoS attacks.

The PCA results are shown in Table 7. The number on the end of each model represents number of features. The original eleven-feature set created

Table 6: Single-attribute model results.

Models	Type 1	Type 2	Prec.	Recall	F-Meas
RF5	9444	700	0.708	0.970	0.818
RF7	9487	714	0.707	0.970	0.818
RF11	9446	706	0.708	0.970	0.818
C4.5N5	9486	715	0.707	0.970	0.818
C4.5N7	9463	709	0.707	0.970	0.818
C4.5N11	9470	709	0.707	0.970	0.818
5NN5	9523	742	0.706	0.969	0.816
5NN7	9540	749	0.705	0.968	0.816
5NN11	9519	748	0.706	0.968	0.816
C4.5D5	9487	722	0.707	0.969	0.817
C4.5D7	9498	720	0.707	0.969	0.818
C4.5D11	9478	720	0.707	0.969	0.818

Table 7: PCA results.

Models	Type 1	Type 2	Prec.	Recall	F-Meas
RF31	9457	715	0.707	0.970	0.818
RF32	9469	743	0.707	0.968	0.817
RF42	9457	733	0.707	0.969	0.818
C4.5N31	10312	59	0.695	0.997	0.819
C4.5N32	9508	751	0.706	0.968	0.816
C4.5N42	9502	739	0.706	0.969	0.817
5NN31	9523	748	0.706	0.968	0.816
5NN32	9558	766	0.705	0.968	0.815
5NN42	9534	770	0.705	0.967	0.816
C4.5D31	10306	82	0.695	0.997	0.819
C4.5D32	9513	765	0.706	0.968	0.816
C4.5D42	9495	749	0.706	0.968	0.817

the forty-two feature set when we applied the PCA filter. We then took the top performing feature sets from the CFS, Consistency, and single-attribute models and applied the PCA filter to them. Two feature selection subsets were selected using five and seven features. The five-feature subset is Bytes, Flags, InitialFlags, BytesPerPacket, and Durmsec. We then applied the PCA filter, generating thirty-one features. The seven-feature set consisting of Packets, Bytes, Flags, Packets/Sec, Bytes/Sec, Bytes/Packet, and Durmsec generated thirty-two features after the PCA filter was applied.

The forty-two feature set had the best results with RF. It had the lowest value for Type 1 and Type 2 errors with 9,457 and 733 respectively. It also had a positive F-measure of 0.818. RF had the best performance compared to C4.5N, C4.5D, and 5NN. The thirty-one feature set produced some interesting results. C4.5D and C4.5N had low Type 2 errors of 82 and 59 respectively, but very high Type 1 errors of 10,306 and 10,312. F-measure scores for the four classifiers had favorable results. Though C4.5N and C4.5D had lower Type 2 errors, RF had less total overall errors (Type 1 9,457, Type 2 715) because of the high number of Type 1 errors with C4.5N and C4.5D (10,312 and 10,306 respectively).

Finally, results from the thirty-two feature set overall were the least favorable in reference to Type 2 errors as compared to feature sets forty-two and thirty-one. Type 1 and Type 2 errors were, on average, higher with all four of our models obtaining adequate F-measure scores compared to the original eleven-feature sets. The exception was the thirty-one feature set, with models C4.5N and C4.5D. C4.5N had the lowest Type 2 errors (59) of all the models we analyzed, but the most Type 1 errors.

The idea of PCA is to use a special coordinate system that depends on the cloud of points as shown in Figure 7. Using Python, we placed the first axis in the direction of greatest variance of the points to maximize the variance along that axis. The second axis is perpendicular to it. In two dimensions there is no choice and its direction is determined by the first axis, but in three dimensions

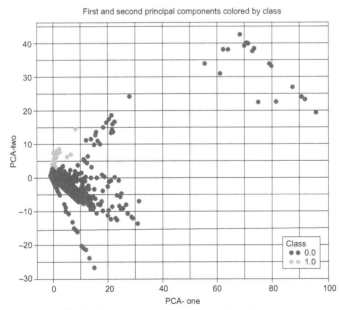

Fig. 7: First and second component colored by class.

it can lie anywhere in the plane perpendicular to the first axis, and in higher dimensions there is even more choice, though it is always constrained to be perpendicular to the first axis. The black areas represent the normal class and the blue are the attacks.

5.2 Learner Results

For each classifier, four runs of five-fold cross-validation are applied, producing 20 AUC values for each classifier. The average AUC values of the four resulting runs and their standard deviations are shown in Table 10.

Overall, the resulting values indicate that six out of eight of our predictive models perform very well at detecting Slow Read attacks, which demonstrates that the features extracted from the network flow can sufficiently distinguish between normal and attack traffic. The classifiers provide reliable results indicating that Netflow features are discriminative enough for detection of Slow read attacks. AUC results for our classifiers show that RF achieved the best performance of 0.96755, but had a higher standard deviation than the other classifiers. The C4.5N, 5NN, and C4.5D learners performed nearly as well as RF, having AUC values of 0.96724, 0.96690 and 0.96620, respectively. These values are only marginally less than those of RF. JRip produced the highest standard deviation, and NB had the lowest AUC.

When evaluating both C4.5 trees, each tree structure selected the Flags feature at the first level of the tree. One of the critical characteristics of a successful Slow Read attack is the flags. There are nine TCP flags shown in Table 8 that make up the attributes for session and initial flags. Here an initial flag value of "S" corresponds to potential attack instances. The value of "S" stands for "SYN," which indicates that a TCP connection was initiated. The value of "P" stands for "PSH" and is like the "URG" flag and tells the receiver

Table 8: TCP flags.

TCP Flag	Description
SYN	Synchronize sequence numbers on new connections.
ACK	Acknowledge the successful receipt of a packet.
FIN	Finished send more data from the sender after a connection is closed.
URG	Process the urgent packets before processing all other packets.
PSH	Process these packets as received instead of buffering them.
RST	Gets sent from the receiver to the sender when a packet is sent to a particular host that was not expecting it.
ECE	Indicates if the TCP peer is ECN (Explicit Congestion Notification) capable.
CWR	Congestion Window Reduced, indicates it received a packet with the ECE flag set.
NS	Nonce Sum protects against accidental malicious concealment of packets from the sender.

to process these packets as they are received instead of buffering them. The value of "A" stands for "ACK" and acknowledges the successful receipt of a packet. The value of "F" stands for "FIN" and represents when a sender is finished sending more data after a connection is closed. The value of "R" stands for "RST" and gets sent from the receiver to the sender when a packet is sent to a particular host that was not expecting it.

JRip and C4.5D produced several rules each. The single consistent behavior between both is in regards to the "flags" attribute. When this attribute has a reset (R) value, there is an attack. Another JRip rule detects an attack when flags have an SPA (SYN, PUSH, ACK) combination, bytes are greater than 512 and the duration in milliseconds is greater than 373. C4.5D uses the FRSPA (FIN, RST, SYN, PSH, ACK) flag along with a combination of more than five packets, bytes less than 770 and a packets per second value less than or equal to 18,633. Our overall analysis allows us to see which features proved most beneficial in the detection of Slow Read attacks.

Figures 8 through 11 illustrate four decision tree examples using the C4.5D model. When the C4.5D model is presented with a flag combination of SPA in Figure 8, we can clearly distinguish discerning behaviors between normal and attack flows. Though the packets per second (pps) may be the same, the attack flow is detected if duration is less than or equal to 31.80 seconds, bytes are less than 32,321, duration greater than 0.374 milliseconds and packets greater than 5. Examining the "FRSPA" flag combination in Figure 9, this combination shows that it can predict an attack if bytes are less than or equal to 770 and the pps is greater than 9,308, but less than or equal to 18,633. When the combination of "FSPA" flags, seen in Figure 10, is present and the duration is greater than 907 seconds, then the instance is labeled as an attack. Figure 11 shows that when the flags are equal to "R", an attack is detected.

The algorithm used for C4.5 decision trees divides the samples into two or more branches based on the values of one of the features in the data sample. The hierarchy of the top-level branches is an excellent source for identifying discriminate features for our normal and attack data. Flags are one of those discriminate features at the top level of the tree that produced Figures 8 through 11. There are thirty values for Flags and the decision tree can quickly locate the relevant values like SPA, FRSPA, FSPA, and R. From these upper tier values, we can identify other relevant features that make up the attribute sets for these four trees. Lastly, we can observe that these attributes are very similar to most of the feature selection methods we used in this work.

As mentioned previously, we used ANOVA to compare the mean AUC values of Slow Read attack detection among the eight learners to check if differences are statistically significant. ANOVA Table 9 shows an F value of 3,388, and a low p-value less than 2e-16. This indicates the variation of means

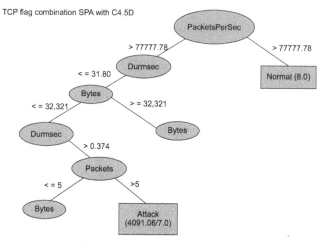

Fig. 8: C4.5D model with SPA flag.

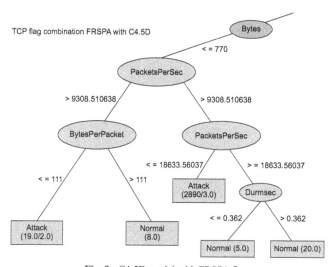

Fig. 9: C4.5D model with FRSPA flag.

among different learners is much more significant than the variation of mean values within each learner at a 95% confidence interval.

Hence, we reject the null hypothesis (i.e., means are not the same). The differences are statistically significant with the relationship between learners and Slow Read attack detection. To determine the statistically significant differences between learners, we conducted a Tukey's HSD post hoc test. The Tukey's test divided our learners into 4 groups based upon their mean AUC values and standard deviations. As shown in Table 10 most groupings resulted

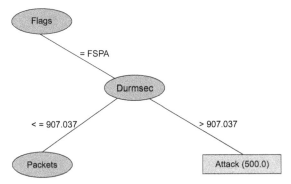

Fig. 10: C4.5D model with FSPA flag.

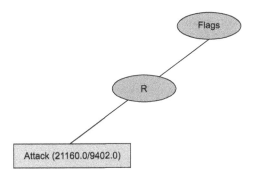

Fig. 11: C4.5D model with the R flag.

Table 9: ANOVA results.

	Df	Sum Sq	Mean Sq	F Value	Pr > F
Learner	7	0.031320	0.004474	3388	Pr < 2e-16
Residuals	24	0.000032	0.000001		

Table 10: Tukey's HSD group results.

Classifier	AUC	AUC StD	Group
RF	0.967554	0.000056	*A*
C4.5N	0.967239	0.000078	*A*
5NN	0.966899	0.000049	*A*
C4.5D	0.966200	0.000038	*A*
MLP	0.950600	0.001600	*B*
JRip	0.947131	0.002600	*C*
SVM	0.892173	0.000504	*D*
NB	0.889400	0.000273	*E*

in fairly close pairings, with the top four learners being in group A and MLP, JRip, NB, and SVM being in their own unique groups.

Table 11 displays how our four group A classifiers performed using feature selection methods against the original eleven features. Acronyms from the table are CFSM (Correlation Feature Selection Method), CSSM (Consistency Subset Selection Method), SASM (Single-Attribute Selection Method), and ORD (Original Dataset). C4.5D performed as well with CFS as it did with its eleven original features. CFS used both Rank-Search and Greedy-Stepwise to produce Bytes, Flags, Initial Flags, Bytes/Packet, and Durmsec for a five-feature set. The data reduction method used by CFS will help C4.5D to decrease the processing time and therefore become more efficient. Feature reduction was also successful with the C4.5N model. The same five-feature set that benefited C4.5D did the same for C4.5N, equaling the same results as those of the eleven-feature set. It also had similar results using the single-attribute method with the seven features; Flags, Bytes, BytesPerPacket, Durmsec, BytesPerSec, PacketsPerSec, and InitialFlags. All four attributes evaluators; GR, Chi-Squared, Symmetric-Uncertainty, and IG had the same false negatives as the original eleven-feature set, but slightly less false positives. CFS and single-attribute methods also provided positive results for 5NN. They both did so, with the five-feature set of Bytes, Flags, InitialFlags,

Table 11: Top classifier results.

Models	Type 1	Type 2	Prec.	Recall	F-Meas
RF5CFSM	9444	700	0.708	0.970	0.818
RF7CSSM	9445	716	0.708	0.970	0.818
RF5SASM	9444	700	0.708	0.970	0.818
RF11ORD	9446	706	0.708	0.970	0.818
C4.5N5CFSM	9470	709	0.707	0.970	0.818
C4.5N7CSSM	9479	716	0.707	0.970	0.818
C4.5N7SASM	9463	709	0.707	0.970	0.818
C4.5N11ORD	9470	709	0.707	0.970	0.818
5NN5CFSM	9423	742	0.706	0.968	0.815
5NN5CSSM	9558	766	0.705	0.968	0.815
5NN5SASM	9523	742	0.706	0.969	0.816
5NN11ORD	9519	748	0.706	0.968	0.816
C4.5D5CFSM	9478	720	0.707	0.969	0.818
C4.5D7CSSM	9581	724	0.707	0.969	0.817
C4.5D5SASM	9487	722	0.707	0.969	0.817
C4.5D11OR	9478	720	0.707	0.969	0.818

BytesPerPacket, and Durmsec showing better performance metrics than the original eleven-feature set. The single-attribute evaluator was GR.

Once again, the five-feature set with CFS and single-attribute methods supplied RF with its best performance. The metrics from this feature set produced the best overall performance from RF for all feature selection methods and classifiers. RF had consistently better metrics in all our experiments. In our opinion, RF is the best classifier to apply to this dataset. The five features of Bytes, Flags, Initial Flags, Bytes/Packet, and Durmsec are the best subset features for discriminating between normal and attack traffic, and providing an efficient processing time.

6. Conclusion

We proposed an approach to successfully detect Slow Read HTTP DoS attacks using machine learning. Our experiments were generated on a live web server that is utilized by faculty and students as well as the public. To improve our machine learning, we implemented various feature selection methods.

By producing our capture, which utilizes real attacks alongside normal data, we better reflect a real-life network environment as compared to existing test-beds. This approach also reinforces the integrity of our data as we are using real-world traffic as opposed to related works which employ simulated traffic. We performed three different variations of a Slow Read attack using the SlowHTTPTest tool to reflect a range of levels. SiLK was used to generate Netflow data using the IPFIX standard. Our experiments show that this approach to Slow Read attack detection produces high AUC values and low false positive and false negative rates. The AUC performance metrics were performed using four runs of stratified five-fold cross-validation.

In addition, we demonstrated that six out of eight learners performed well in the detection of Slow Read attack traffic, with four of them performing significantly better than the rest. As a result, Netflow features have shown that they are able to successfully detect distributed Slow Read HTTP DoS attacks.

The five-feature subset of Bytes, Flags, InitialFlags, BytesPerPacket, and Durmsec performed better than the three or eleven-feature sets. RF and 5NN had lower Type 2 errors using feature selection. C4.5N and C4.5D had the same amount of Type 2 errors with the five and eleven-feature sets. CFS with GS and single-attribute GR had the best results with overall Type 1 and Type 2 errors when we ran our group A classifiers (RF, C4.5N, 5NN, C4.5D). Feature selection was very effective with all four learners and demonstrated with five features more favorable accuracy, predictability, and less chance of overfitting.

Overall, the Consistency Subset Evaluator with RandomSearch performed well with seven features. RF was the best overall learner, using the seven-feature set of Protocol, Bytes, Flags, Packets/Sec, Bytes/Sec, Bytes/Packet, and Durmsec that was generated by the Random Search method. RF again had the best overall performance with the single-attribute method. RF shared the same results as those with CFS which also had the best overall results of all feature selection methods and the original eleven-feature dataset.

PCA results did not fare as well as the other three methods. The thirty-two feature set obtained after PCA filtering was applied produced the best results. Most of the results from our feature selection methods performed as well as the original full feature set of eleven features. CFS accomplished this with five features, Consistency with seven, and the single-attribute method with five. Producing similar results with less features will improve model accuracy and predictability.

Future work will involve collecting traffic for another application layer DDoS attack called a POST attack. We plan to evaluate if Netflow features also provide discriminating detection for other attack variants.

Acknowledgment

We would like to thank the reviewers in the Data Mining and Machine Learning Laboratory at Florida Atlantic University. Additionally, we acknowledge partial support by the NSF(CNS-1427536). Opinions, findings, conclusions, or recommendations in this paper are the authors' and do not reflect the views of the NSF.

References

[1] Are you ready for slow reading. [Online]. Available: http://blog.shekyan.com/2012/01/are-you-ready-for-slow-reading.html.

[2] Dos website using slowhttptest in kali linux â" slowloris, slow http post and slow read attack in one tool. [Online]. Available: https://www.blackmoreops.com/2015/06/07/attack-website-using-slowhttptest-in-kali-linux/.

[3] Probing zero windows. [Online]. Available: https://tools.ietf.org/html/rfc1122#page-92.

[4] Radware's ddos handbook: The ultimate guide to everything you need to know about ddos attacks. [Online]. Available: https://security.radware.com/uploadedfiles/resources_and_content/ddos_handbook/ddos_handbook.pdf.

[5] Silk. [Online]. Available: https://tools.netsa.cert.org/silk/index.html.

[6] Q2 2017 akamai state of the internet/security report. Tech. Rep. [Online]. Available: https://content.akamai.com/us-en-pg9565-q2-17-state-of-the-internet-security-report.html.

[7] Adi, E., Z. Baig and P. Hingston. 2017. Stealthy denial of service (dos) attack modelling and detection for http/2 services. Journal of Network and Computer Applications 91: 1–13. [Online]. Available: http://ww.sciencedirect.com/science/article/pii/S1084804517301637.

[8]   Bauder, R. and T.M. Khoshgoftaar. 2018. Medicare fraud detection using random forest with class imbalanced big data. pp. 80–87. *In*: IEEE International Conference on Information Reuse and Integration (IRI). IEEE.

[9]   Benzel, T., B. Braden, T. Faber, J. Mirkovic, S. Schwab, K. Sollins and J. Wroclawski. 2009. Current developments in deter cybersecurity testbed technology. pp. 57–70. *In*: Conference for Homeland Security, 2009. CATCH'09. Cybersecurity Applications & Technology. IEEE.

[10] Blackert, W., D. Gregg, A. Castner, E. Kyle, R. Hom and R. Jokerst. 2003. Analyzing interaction between distributed denial of service attacks and mitigation technologies. pp. 26–36. *In*: DARPA Information Survivability Conference and Exposition. Proceedings, Vol. 1, IEEE.

[11] Calvert, C., C. Kemp, T.M. Khoshgoftaar and M. Najafabadi. 2017. A framework for capturing http get ddos attacks on a live network environment. pp. 136–142. *In*: International Society of Science and Applied Technologies, ISSAT.

[12] Claise, B. 2004. Cisco systems netflow services export version 9 (rfc 3954). [Online]. Available: https://www.ietf.org/rfc/rfc3954.txt.

[13] Claise, B., B. Trammell, E. Zurich and P. Aitken. 2013. Specification of the ip flow information export (ipfix) protocol for the exchange of flow information (rfc 7011). [Online]. Available: https://tools.ietf.org/search/rfc7011.

[14] Cohen, W.W. 1995. Fast effective rule induction. pp. 115–123. *In*: Twelfth International Conference on Machine Learning. Morgan Kaufmann.

[15] d'Aspremont, A., L.E. Ghaoui, M.I. Jordan and G.R. Lanckriet. 2005. A direct formulation for sparse pca using semidefinite programming. pp. 41–48. *In*: Advances in Neural Information Processing Systems.

[16] Farnaaz, N. and M. Jabbar. 2016. Random forest modeling for network intrusion detection system. Procedia Computer Science 89: 213–217. [Online]. Available: http://www.sciencedirect.com/science/article/pii/S1877050916311127.

[17] Gaonjur, P., N. Tarapore, S. Pukale and M. Dhore. 2008. Using neuro-fuzzy techniques to reduce false alerts in ids. pp. 1–6. *In*: Networks. ICON 2008. 16th IEEE International Conference on, IEEE.

[18] Haddadi, F., J. Morgan, E.G. Filho and A.N. Zincir-Heywood. 2014. Botnet behaviour analysis using ip flows: With http filters using classifiers. pp. 7–12. *In*: 28th International Conference on Advanced Information Networking and Applications Workshops, May 2014.

[19] Hall, M., E. Frank, G. Holmes, B. Pfahringer, P. Reutemann and I.H. Witten. 2009. The weka data mining software: An update. pp. 10–18. *In*: SIGKDD Explor. Newsl, Vol. 11, ACM.

[20] Hares, S. and D. Katz. 1989. Administrative domains and routing domains a model for routing in the internet (rfc 1122). [Online]. Available: https://www.ietf.org/rfc/rfc1136.txt.

[21] Inacio, C.M. and B. Trammell. 2010. Yaf: yet another flowmeter. In Proceedings of LISAâTM10: 24th Large Installation System Administration Conference, p. 107.

[22] Jethanandani, M. 1013. Tcp may keep its offered receive window closed indefinitely (rfc 1122). [Online]. Available: http://www.kb.cert.org/vuls/id/723308.

[23] Li, J., K. Cheng, S. Wang, F. Morstatter, R.P. Trevino, J. Tang and H. Liu. 2017. Feature selection: A data perspective. ACM Computing Surveys (CSUR) 50(6): 94.

[24] Liu, H. and R. Setiono. 1996. A probabilistic approach to feature selection—a filter solution. pp. 319–327. *In*: 13th International Conference on Machine Learning.

[25] Lucio, G.F., M. Paredes-Farrera, E. Jammeh, M. Fleury and M.J. Reed. 2003. Opnet modeler and ns-2: Comparing the accuracy of network simulators for packet-level analysis using a network testbed. WSEAS Transactions on Computers 2(3): 700–707.

[26] Maaten, L.v.d. and G. Hinton. 2008. Visualizing data using t-sne. Journal of Machine Learning Research 9(Nov.): 2579–2605.

[27] McCanne, S., S. Floyd, K. Fall and K. Varadhan. 1995. The network simulator ns2 (1995) the vint project, Available for download at http://www.isi.edu/nsnam/ns.

[28] Mirkovic, J., S. Fahmy, P. Reiher and R.K. Thomas. 2009. How to test dos defenses. pp. 103–117. *In*: Conference for Homeland Security, 2009. CATCH'09. Cybersecurity Applications & Technology, IEEE.

[29] Moore, A.W. and D. Zuev. 2005. Internet traffic classification using bayesian analysis techniques. SIGMETRICS Perform. Eval. Rev. 33(1): 50–60, Jun. 2005. [Online]. Available: http://doi.acm.org/10.1145/1071690.1064220.

[30] Mukherjee, S. and N. Sharma. 2012. Intrusion detection using naive bayes classifier with feature reduction. Procedia Technology 4: 119–128.

[31] Najafabadi, M.M., T.M. Khoshgoftaar, C. Calvert and C. Kemp. 2015. Detection of ssh brute force attacks using aggregated netflow data. pp. 283–288. *In*: Machine Learning and Applications (ICMLA), 2015 IEEE 14th International Conference on, IEEE.

[32] Najafabadi, M.M., T.M. Khoshgoftaar, C. Calvert and C. Kemp. 2017. User behavior anomaly detection for application layer ddos attacks. pp. 154–161. *In*: Information Reuse and Integration (IRI), 2017 IEEE International Conference on, IEEE.

[33] Najafabadi, M.M., T.M. Khoshgoftaar, C. Calvert and C. Kemp. 2017. A text mining approach for anomaly detection in application layer ddos attacks. pp. 312–317. *In*: FLAIRS Conference.

[34] Najafabadi, M.M., T.M. Khoshgoftaar and A. Napolitano. 2016. Detecting network attacks based on behavioral commonalities. International Journal of Reliability, Quality and Safety Engineering 23(01): 1650005.

[35] Najafabadi, M.M., T.M. Khoshgoftaar and N. Seliya. 2016. Evaluating feature selection methods for network intrusion detection with kyoto data. International Journal of Reliability, Quality and Safety Engineering 23(01): 1650001.

[36] Ndibwile, J.D., A. Govardhan, K. Okada and Y. Kadobayashi. 2015. Web server protection against application layer ddos attacks using machine learning and traffic authentication. pp. 261–267. *In*: Computer Software and Applications Conference (COMPSAC), 2015 IEEE 39th Annual, Vol. 3, IEEE.

[37] Panda, M., A. Abraham and M.R. Patra. 2015. Hybrid intelligent systems for detecting network intrusions. Security and Communication Networks 8(16): 2741–2749. [Online]. Available: http://dx.doi.org/10.1002/sec.592.

[38] Park, J., K. Iwai, H. Tanak and T. Kurokawa. 2014. Analysis of slow read dos attack and countermeasures. pp. 37–49. *In*: The International Conference on Cyber-Crime Investigation and Cyber Security (ICCICS2014). The Society of Digital Information and Wireless Communication.

[39] Peterson, L., A. Bavier, M.E. Fiuczynski and S. Muir. 2006. Experiences building planetlab. pp. 351–366. *In*: Proceedings of the 7th Symposium on Operating Systems Design and Implementation. USENIX Association.

[40] Raschka, S. 2014. Naive bayes and text classification i.

[41] Raschka, S. and V. Mirjalili. 2017. Python Machine Learning, 2nd Ed. Packt Publishing.

[42] Roweis, S.T., L.K. Saul and G.E. Hinton. 2002. Global coordination of local linear models. pp. 889–896. *In*: Advances in Neural Information Processing Systems.

[43] Sachdeva, M., G. Singh, K. Kumar and K. Singh. 2010. Measuring impact of ddos attacks on web services.

[44] Sanders, C. and J. Smith. 2013. Applied Network Security Monitoring: Collection, Detection, and Analysis. Elsevier.

[45] Saravanan, R., S. Shanmuganathan and Y. Palanichamy. 2016. Behavior-based detection of application layer distributed denial of service attacks during flash events. Turkish Journal of Electrical Engineering & Computer Sciences 24(2): 510–523.

[46] Seliya, N., T.M. Khoshgoftaar and J. Van Hulse. 2009. A study on the relationships of classifier performance metrics. pp. 59–66. *In*: Tools with Artificial Intelligence, 2009. ICTAI'09. 21st International Conference on, IEEE.

[47] Sharma, A.K. and P.S. Parihar. 2013. An effective dos prevention system to analysis and prediction of network traffic using support vector machine learning. International Journal of Application or Innovation in Engineering & Management 2(7): 249–256.

[48] Singh, K.J. and T. De. 2015. An approach of ddos attack detection using classifiers. pp. 429–437. *In*: Shetty, N.R., N. Prasad and N. Nalini (eds.). Emerging Research in Computing, Information, Communication and Applications, New Delhi: Springer India.

[49] Witten, I.H., E. Frank, M.A. Hall and C.J. Pal. 2016. Data Mining: Practical Machine Learning Tools and Techniques. Morgan Kaufmann.

[50] Wueest, C. 2014. Security response: The continued rise of ddos attacks. [Online]. Available: http://www.symantec.com/content/en/us/enterprise/media/securityresponse/whitepapers/the-continued-rise-of-ddos-attacks.pdf.

[51] Yang, J., A. Tiyyagura, F. Chen and V. Honavar. 1999. Feature subset selection for rule induction using ripper, 03.

[52] Zeifman, I. 2017. Global ddos threat landscape q3 2017. [Online]. Available: https://www.incapsula.com/ddos-report/ddos-report-q3-2017.html.

[53] Zhang, J., C. Chen, Y. Xiang, W. Zhou and Y. Xiang. 2013. Internet traffic classification by aggregating correlated naive bayes predictions. IEEE Transactions on Information Forensics and Security 8(1): 5–15.

Chapter 9

Predictive Analysis of Server Log Data for Forecasting Events

Reeta Suman,[1,*] Behrouz Far,[1,*] Emad A Mohammed,[2] Ashok Nair[3] and Sanaz Janbakhsh[3]

1. Introduction

IT System log server keeps the record of all the activities and relevant information about the servers. The data analysts and maintenance team manually examine these logs to identify the behavior of the system prior to the error logs. The log error messages do not occur in the same pattern and sometimes go through different stages and accumulate extra information, which makes the time-series prediction task more challenging; therefore, it is critical to choose the accurate predictive model to predict the future events for proactive actions. The previous paper [17], explains the partial architecture of the existing system as follows, the monitoring server called hawk server, generates the monitoring logs, the proxy server named Nginx generates the proxy server logs, and admin server generates the admin server logs. All the logs collectively stored in the log file database, which makes it harder to analyze the error logs. whenever the error happens or if the system services shut down data analysts explore the log data and gain the insight of system and find out the related issues to bring the system back to up and running

---

[1] Dept. of Electrical and, Computer Engineering, University of Calgary.
[2] Dept. of Software Engineering, Lakehead University.
[3] Analytics and Integration, Team, City of Calgary.
  Emails: emohamme@lakeheadu.ca; Ashok.Nair@calgary.ca; sajan@deloitte.ca
* Corresponding authors: rsuman@ucalgary.ca; far@ucalgary.ca

condition [1], [2]. In general, the analytics team spends a significant amount of their time and efforts to analyze the logs and searches the related logs with the keyword search. In order to understand the unexpected behavior of the system, one must understand the log file characteristics. The log files records have the following information related to the state of the system:

1) The log files have all the information about what had happened just before the system went down.
2) Error logs are an essential source of diagnostic and for proactive error handling purpose [20].
3) Unrelated errors are identified in the server logs due to other application or services of another system such as monitor server logs have the error from the proxy server.
4) Historical errors also help to understand the behavior of the system, how to repair the system and how to mitigate the issues.
5) The log files also contain information about the causes of specific issues of the system.

The log files data contain the time-stamp and long error message which include lot of information. We extract all the important information in the data processing section. In recent years, several tools have been developed for analyzing logs, clustering the logs based on IP-address and creating a sequence of logs. For example, [1], [2], [4] propose the methodology of generating a sequence of logs. In order to extract the sequence out of the data, we assign the Ei's to the error messages, and Ei's are the event label which is assigned based on error message and origin of the error. We remove duplicates based on timestamp, which helps to reduce the number of logs and also reduces the human efforts of exploring long list of data. Here choosing the window size is a key parameter for extracting the desired results from the log data, and the size of window also depends on the data. We experimented with different window sizes and found that if we take one hour window then the extracted sequence gets larger and frequency of sequence decreases and if we take 15-minute window the sequence is really short and if we take 30-minute window, it, shows the seasonality in the data, and we also used moving window of 30 minutes to extract the short sequence of errors and analyze data for a day.

The main focus of this study is to predict the accurate event in time-series and our approach can anticipate the upcoming event and time of the event based on the historical data and we also want to measure the accuracy of model between statistical model and machine learning model for the server log analysis. We are exploring the LSTM as machine learning algorithm, Holt-Winters, and ARIMA as statistical model and we have implemented the LSTM, ARIMA and Holt-Winters models in accordance to server log analysis. Our approach provides the visualization of predicted events to the data analysts, so that user can take proactive steps for error handling. We are finding the pattern in the sequence of logs, which is helping us to find the subsequent events are happening on the server and help us to predict the error and error of time in future through LSTM, Holt-winters, and ARIMA. The key contributions of this paper are:

1) Investigating the forecasting techniques for time-series and predicting future events for server logs.
2) Comparing the performance of LSTM, Holt-Winters and ARIMA algorithms, and which one has higher accuracy results for predicting sequence of events.
3) Provide visualization of predicted events for server log time-series.
4) Implementation of LSTM machine learning model, Holt-Winters, and ARIMA statistical model.

The chapter is organized as follows. We have presented Section 2, as related work. Section 3 is a case study and our motivation. In Section 4, we have our proposed approach, and in Section 5, we have results of our approach. Section 6, Conclusion, Section 7 is limitations, and future work of our approach and references are in last section.

2. Related Work

This section is to provide the review of related works. Jiang et al. [1] proposed an automated approach which examines the logs, recognize the internal structure of log lines and then convert them into related execution events. Once the developer recovered the log structure, analyzing log behavior become easy. The abstraction of the logs reduces the volume of data to examine, and then the log lines get converted to the execution events by anonymized step for recognizing dynamic token replace with the generic token. The log abstraction technique applied to the source code of the program, and they developed the approach to recognizing the structure of log message and this approach perform well on large log files. We found that this method is only applicable to the applications if the source code of an application is available to the analyst, which is not possible in our case. Lin et al. [2] have utilized the knowledge base to check the sequence of a log that existed before. They used the log sequences for manual examination and then applied mitigation actions. They have applied their approach to Microsoft services clustering and validated it; however, in our existing systems at City of Calgary log clustering is not sufficient for the unstructured data of the log messages because each server and service create different types of logs which may lead to false positives. This methodology cannot be generalized for generic log analysis.

Zhen et al. [4] worked on log Abstraction, log linking, simplifying sequences and generating execution reports of the simple apps worked on abstracting execution logs from big data Hadoop based cloud environment, and they recover the execution sequences and generating execution reports of the simple apps by comparing the sequences between pseudo cloud environment and Google cloud environment. Their approach uncovers the behavior of big data application platform behavior and doesn't validate for developing and testing applications. They also injected the deployment faults to verify their results and prove that their approach significantly reduces the verification efforts. They reduced the logs to a too small size that the manual inspection can be done on the sequence of logs. They have experimented only on the local platform logs; whereas we are concentrating on the server logs.

Xu et al. [5] proposed methodology to detect system issues by mining the console logs. Using source code helped to understand the structure of console logs. They have designed an online log parsing technique and extracted powerful features to reveal the information, and then applied principal component analysis for the anomaly detection. The log parsing methodology is based on the system source code however this approach cannot be applied if we do not have access to source code. Shilin He et al. [6], worked on state of the art for anomaly detection, which consists of log collection, log parsing, and feature extraction then used anomaly detection. They also explored the detailed review of supervised and unsupervised techniques. However, the algorithm for anomaly detection cannot be generalized because the experiment was done under lab settings.

Bovenzi et al. [8] worked on anomaly detection and explored the methodology which can be fitted at an operating system level. The algorithm diagnoses the activities of the software system of the operating system, and this methodology can only apply to the operating system. Chandola et al. [11] explored the anomaly detection methodology in a survey. They stated that sometime the events happen at a specific point of time, which is called contextual anomaly and some time the events occur in an unordered list which is called the collective anomaly.

Data mining of logs can be divided into the following categories as descriptive mining, prescriptive mining and predictive mining. Descriptive data mining usually based on describing the domain and predictive log data mining based on making predictions and prescribe the solutions. Hay et al. [28] finding sequential patterns and finding an association between the events are part of the descriptive analysis and predicting the events based on historical data is partially related to anomaly detection. The objective of predictive analysis in this chapter is to estimate the unknown future events and time of the event. In order to estimate, we have explored the predictive models such as statistical models and machine learning algorithms. Salfner

Table 1: Review of three models for predictive analysis.

Model	Pros	Cons
Holt-Winters	Ability to handle trend, and seasonality	Narrow confidence interval
ARIMA	Unbiased forecast Easy to derive confidence interval for forecast	Require more data Hard to automate
LSTM	Ability to handle complex nonlinear patterns Easily automate	Difficult to derive confidence intervals for forecast Require more data

et al. [20] worked on pre-processing of error log data for accurate failure prediction. The information in the log files are essential and valuable, and they experimented the log clustering and filtering technique in order to get the precise information, and they have explored online failure statistical technique based on hidden semi-Markov predictor (HSMM) and appropriate pre-processing of data. HSMM is efficient learning algorithm for raw sequence and discovering the patterns however it is unable to express the dependencies and correlation between the patterns. Namin et al. [21] explored the forecasting technique of ARIMA and LSTM and compared the accuracy of the results. They experimented on financial and economic data and used Epoch technique using different size of windows; however, the epoch sizing did not help to improve the results. Chatfield and Yar [23] used Holt-Winters method to predict the future values based on triple exponential smoothing methodology if the data have trend and seasonality. Based on these studies, we apply the predictive models to our data set to find out appropriate models. In Table 1 we have described the review of statistical and machine learning algorithm, which is explained in further sections of the chapter.

3. Case Study and Motivation

In the modern era of technology, part of the success of companies depend on the reliability of their servers and IT-services and in case of service shutdown, data analysts analyze the log data. In our approach, we are creating the sequence of logs and predicting the event which further helps the analytics team to examine the logs and they can mitigate the issue before it happens. Abnormal behavior can also be detected if expected event does not occur, and the data analyst can also identify an unusual pattern that does not fit the predictable response [11], [1]. We have analyzed the log data manually and observed the patterns of errors as shown in Figure 1. The patterns of events show that some of the events occur in some desired fashion and we can predict the future event. We are displaying Ei's events in some pattern over the period in Figure 1. The event Ei's are assigned based on error text and category of the

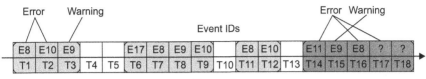

Fig. 1: The manual observation of the events generated by the monitoring server and shows the sequence of errors occurs at a particular point in time and in the similar fashion, which can help us to predict the specific event for next period.

error. In this chapter of the book, we are exploring the sequence of errors and analyzing the patterns of events in order to predict the future event. Abnormal behavior of logs can also be detected if the events do not confront the specific pattern. We are also considering unusual contextual behavior and collective abnormal behavior. In Figure 1 we show the Event E8 and E10 always happen together, and the E9 raises the warning for the user. The dataset of logs used for the experiments is extracted from the IT department of the City of Calgary under the analytics and integration team. The log file contains the essential information about the system and server such as timestamp, detail description of the error, error number, the application name related to failure, and server address. We have access to the error logs of servers and application which we are using to validate our approach and analyzing the results.

We gather information about the servers, applications, messaging service, and databases, which generates log files. We want to gain the insight of those logs by studying them and finding critical patterns of the loglines, which leads to complete system or server failure. Our approach reduces the time and efforts of data analyst and provides a solution to analyze the logs accurately, which make them more productive for analyzing the logs. We are also reducing the number of logs to analyze through creating events from the logs, for example we took sample data of one day from monitoring server, which contains 1733 log lines after generating events out of the log, it reduces to 89 events and 23 sequence of events, which reduces the efforts by 94%. Once we learn the sequence of events and pattern, we explore it to predict the future events or sequence of events.

4. Our Proposed Approach

Time-series analysis is the area of information in which we create the model based on observation, historical data, and predict future values. Forecasting time-series is applied to many areas, therefore, it is critical to improving the existing forecasting model with new emerging techniques such as deep learning. In this chapter of the book, we have expanded the approach from IRI 2018 proceeding [17] to the predictive analysis. Figure 2 is the extended

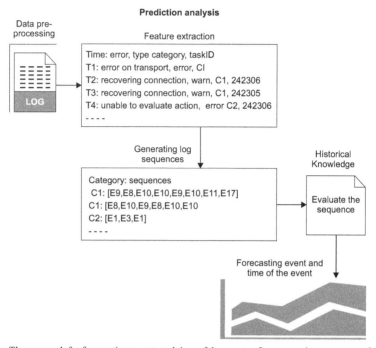

Fig. 2: The approach for forecasting events and time of the events after generating sequence of events.

version of our approach. We started our approach with data gathering and lead to forecasting techniques. The framework of our approach is divided into three main phases; The first two phases are the repetition from the previous research paper in which we discuss data pre-processing and feature extraction. The second phase is converting the error logs into a sequence of log lines or events, and the third phase forecasting the events and time of the events from the sequence of events.

Time-series forecasting involves mathematical processes, and pattern search, which plays a significant role in the data analysis [24] such as ARIMA and Holt-Winters models, which have long been in use for stock market time-series forecasting. LSTM is a modern and new emerging technology in deep learning to address the prediction issues. The Auto Regressive Integral moving average (ARIMA) is a method in which a regressive model confirms the dependencies between the observation; integration means measuring differences of observation and moving average lag of the existed observations. In this chapter, we are focusing on traditional and modern techniques for time-series forecasting and evaluating accuracy for log data prediction. We have applied ARIMA, Holt-Winters and LSTM methodology for the forecasting of the events of the server.

4.1 Data Preprocessing and Feature Extraction

Data preprocessing is critical for data mining and includes data cleaning, integration, transformation, data selection, and data reduction. Log data files are full of text, including important error messages, symbols and blank lines. Whenever; the new event happens, or the date gets changed on the server, some blank lines and system software information get generated, which is not relevant to the log analysis, therefore removing blank lines, and symbols is important step for cleaning and preprocessing the data. We have converted the textual data to the data-frames or tables and converted timestamp text to date time format since it's a manageable way to store the information about the date. The critical information gets extracted from the textual data of logs, and we manage the missing values, smooth out noisy data, and we transform the time to minutes. We extract the error messages from the text of the error and remove the irrelevant information by filtering out exceptional handling log data.

We have reduced the features by selecting the important attributes from the data as shown in the Table 2 and split the data and kept 70% data for the training and 30% data for the testing.

Table 2: After data processing, the server log information is stored into data-frames for the further analysis steps and shows the text data has been cleaned and preprocessed.

Date	Time	Error	Type	Category	TaskID
June 19	3:00:50 PM	Error on transport	Error	HWKCON	242304
June 19	3:00:50 PM	Recovering connection	Warn	HWKCON	242305
June 19	3:01:00 PM	Recovering connection	Warn	HWKCON	242305
June 19	3:01:00 PM	Session is closed	Error	HWKCON	241809
June 19	3:01:02 PM	Recovering connection	Warn	HWKMAG	242305
June 19	3:01:07 PM	Unable to evaluate action	Error	HWKRBE	41415

4.2 Generating Log Sequences

We have extracted the number of features from the logs, however, and considered only relevant information of error messages. The server logs include the date and time, error message, type of event, category, and TaskID of the log message, and have some constant and dynamic part of the error message. We have used the algorithm from IRI 2018 proceeding [17] for assigning the event ids to the error messages which further get converted to event sequences. We assign the Ei's to error message based on task id's and link them based on categories. The events E1, E4, E5, E5, E1, E1, E7, E7 are under HWKRBE category, and they get generated at the same time and are

Table 3: The sequence of events in fixed window technique. The grouping of events has been done to create the sequence and removing the duplicate events in the row.

Date	Time	Error	Type	Category	TaskID	EIs
June 18	3:06:50	Recovering connection	Warn	HWKCON	242305	E9
June 18	3:06:50	Error on transport	Error	HWKCON	242304	E8
June 18	3:07:00	Exception sending msg	Error	HWKCON	241809	E10
June 18	3:07:01	Exception sending msg	Error	HWKCON	241809	E11
June 18	3:08:00	Recovering Connection	Warn	HWKCON	242305	E9

linked together with the same category type. The events E9, E8, E10, E11, E9, are under HWKCON, and they are linked together. Linking step accumulates the duplicate for the event ids to the specific category in the fixed windows. The sequence of events has been created by grouping the repeated events to one event and removing the duplicate events [4]. Table 3 is an example of the sequences of logs, with timestamp, error messages, and TaskID; which are linked together since they have similar categories or type of event.

4.3 Predicting and Visualizing Log Events

We have explored LSTM, ARIMA, and Holt-winters algorithms for the experimentation purpose. LSTM (Neural network) is the artificial machine learning model; however, ARIMA (Autoregressive), and holt-winters (exponential smoothing) are statistical models for the predictive analysis. The algorithm predicts the future event and time of the event, based on historical data. One of the challenges for using the machine learning models and statistical model that, the data must be numerical; therefore in order to predict event and time of the event, we convert time into minutes and categorical data such as event Ei's to numeric values, in order to input the data to statistical model and machine learning models.

4.4 Predicting Future events by LSTM

Long Short-term memory (LSTM) is the extension of recurrent neural networks, which has the capability of remembering long sequences and the historical values. The model can be trained using backpropagation through time which makes them suitable for time-series prediction. LSTM creates the knowledge base from the historical data and very powerful in handling the dependencies between the inputs and smart enough to determine whether to hold the information or forgets the information. In this chapter of the book, we have implemented LSTM model in order to forecast the events and time of the event. The model uses historical data to identifies the existing data patterns and use them to predict the future event and time.

Implementation of LSTM model

Input: Series of events in dataframe
Output: Predicted Events and Root Mean Square
 # Input libraries
 # fix the random seed for reproducing results
1.   Set random. seed (number)
 # convert dataset to supervised data to feed in the model
Procedure: timeseries (data, shift_data ← 1)
2.   data ← dataframe(data)
3.   column ← shit(i) for I in range (1, shift_data+1)
4.   column.append(data)
5.   data ← pd.concat (column, axis ← 1)
6.   return data
 # normalize the data
7.   scaler ← MinMaxScaler (range ← (−1,1)
8.   dataset ← scaler.fit (dataset)
 # transform data
9.   dataset ← dataset.reshape(dataset.shape[0], dataset.shape[1])
10.   dataset ← scaler.transform(Dataset)
 # Fit an LSTM model to training data
11. **Procedure**: fit_lstm (dataset, batch, epoch, neurons)
12.   x ← dataset
13.   y ← dataset − x
14.   model ← Sequential ()
15.   model.add(LSTM(neurons, batch))
16.   model.add (Dense (1))
17   model.compile(loss ←"Mean_square_Error",
                optimizer ← adam)
18.   forloop I in range(epoch)
19.   model.fit (x,y, epochs ← 1, batch)
20.   return model
 # prediction of the events
21. Procedure forecast (model, x)
22.   yhat ← model.predict(x)
23.   return yhat

The Keras library are used for the LSTM model; however, the data preparation is needed for the machine learning techniques. In order to measure the accuracy of the LSTM prediction model, and we have used RMSE Root Mean Square Error.

4.5 Predicting Future Events by ARIMA

Autoregressive Integrated Moving Average (ARIMA) combines the regressive process and moving average method and construct the composite model [21, 22]. An autoregressive model is based on dependencies between observations of historical data; integrated means taking the difference between observation and the previous observation, to make the time series stationary; and moving average take the dependencies between observation and residual error. The mathematical formulation of ARIMA (p, d, q) model had three parameters number of lags, number of times and size of the windows as p, d, and q. In order to use model, we determine the p, d, q which are integers greater or equal to zero and we need to make the time-series data stationary then construct the ARIMA model to make a prediction and find the correlation of the time-series itself. The integer d determines the differencing and if $d = 0$ then results depends on p and q. We transform a series to log series to make it stationary and then take the difference of log_transform and log_transform_ shift by 1. We create autocorrelation factor and partial autocorrelation factor plots to identify the patterns.

Implementation of ARIMA

 # ARIMA Model with rolling window
Input: series in dataframe
Output: forecasted events and root mean square
 # load libraries
 # test the stationarity of the dataset
1. **Procedure**: test_stationarity(dataset)
2.   movingAVG ← dataset_log.rolling(window ← 30).mean()
3.   movingstd ← dataset_log.rolling (window ← 30).std()
4.   data_shift ← dataset_log-dataset_log.shift()
5.   lag_acf ← acf(data_shift, nlags ← 30)
6.   lag_pacf ← pacf(data_shift, nlags ← 30, method = 'ols')
7. **# plotong the acf and pacf**
8.   plt.plot(lag_acf)
9.   plt.plot(lag_pacf)
10.   model ← ARIMA (dataset_log, order ← (p,d,q))
11.   result ← model.fit(disp ← −1)
12.   model_arima_fit ← model_arima.fit()
 # Prediction of the events
13.   Prediction ← series (model_arima_fit.fittedvalues, copy ← true)
14.   rms ← sqrt (mean_square_error(dataset_log, prediction)

4.6 Predicting Future Events by Holt-Winters

Holt-Winters is an exponential smoothing statistical model for the time series and is suitable for the data, with seasonality and trend, where the seasonality can be defined as the tendency, which shows the behavioral pattern [28]. Holt winters forecasting is a mathematical model, which predicts the behavior of the time-series and also called triple exponential smoothing. Holt winters also contain the main three parameters (α,β,γ) and predict current or future value based on the parameters. The Alpha, beta, and gamma optional parameters determine the stability of the forecast and values range from 1 and 0. Alpha determines the weighted average of the points; beta determines the slope between consecutive points and gamma determines the seasonality of the series and by default, these values are set to 1 and the seasonality is a repetition of data in the fixed length of time.

Implementation of Holt-Winters

```
# Holt Winters model for the predications
Input: series in dataframe
Output: forecasted events and root mean square
1. # Load the exponential smoothing libraries
2. dataset_log = np.log(dataset)
# triple exponential smoothing multiplicative/additive
# here x is data, m is period, fc is forecast
# Model ← Multiplicative(x,m,fc,alpha,beta,gamma)
3. y_hat_avg ← dataset.copy()
4. fit ← ExponentialSmoothing(dataset, seasonal_periods=5,
trend=mul, seasonal ← 'add').fit()
5. y_hat_avg['Holt_Winter'] ← fit.forecast(len(dataset)
6. rms ← sqrt(mean_squared_error(dataset),
y_hat_avg.Holt_Winter))
```

We experiment with data collected from integration and analytics team from the City of Calgary, and all the log files contain error logs, info logs, and warning logs. The server error logs include timestamp and error description with other information. We have used three types of Log files from the server, such as Proxy web server, monitoring server, and admin server. Complete logs show the data for three months and sample logs have one-day log data. In order to have a proof of concept experiment setting, we have done our experiments on available data captured over three months of period. The output of the extracted sequence of the events raises a question that, does our approach reduce the efforts, and which method of forecasting is appropriate for log analysis?

5. Results

5.1 Results for Generating Sequences

The approach for creating sequences is helping data analyst for analyzing log data to make an informed decision. We have set up the experiment environment on the local computer. The features are fed into the model for generating the sequences. These sequences have been generated after removing all the duplicates of entries. In reference to the time of the event gives the opportunity to data analyst for the detailed review of the log events happening at a particular point of time. The data analyst or user has to look for only a few log events instead of looking complete log lines. In the previous proceeding paper [17], we mentioned the significant reduction of efforts approximately 91.6% and the analyst took less than 5 min to analyze the logs. The sequence of events shows the number of sequences found in the sample log data and Table 4 we have shown the event sequences found in the sample data, and the total number of log lines in the sample file is 1733. Once we removed the info lines from the logs, our log events reduced significantly; even data analysts are able to evaluate the results manually. Suzgun et al. [16] worked on the sequence prediction with short and long sequences. In our case the sequence of events is not larger than size 10, therefore the LSTM model will not face NP-hard issue.

Table 4: The sequence of events in the 30-minute fixed window, which contain errors and warnings. The grouping of events is based on categories, and we have run the algorithm to count the existed sequences from monitoring server logs.

Category	Frequent Sequences	Counts
HWKCON	E9, E8, E10, E10, E9, E10, E11, E17	12
HWKCON	E8, E10, E9, E8, E10, E10	4
HWKRBE	E1, E3, E1	5
HWKRBE	E1, E4, E5, E1, E7	2

5.2 Results for Predictive Analysis

Root Mean Square Error of our Results

Root-Mean-Square-Error (RMSE) is used to measure the accuracy of the model by taking the difference between predicted, and actual values. We have used RMSE with predictive models to evaluate the accuracy of the predictions of the events and time of the event. Alexei et al. [15] identify RMSE is scale dependent measure and useful for measuring the accuracy of models.

$$RMSE = \sqrt{\frac{1}{N}\sum_{i=1}^{N}(a_i - \hat{a}_i)^2}$$

The formula above has N, which is the total number of observations and (a_i) represent the actual values and ($\hat{a}_i$) represent the predictive values. Mugume et al. [14] investigated the performance of the numerical model by RMSE-Root Mean Square Error, MAE-Mean Absolute Error, ME-Mean Error, and STM-Sign test to analyze the bias of numerical method and mentioned that RMSE is good criteria to classify the accuracy of the model and low index value means higher accuracy.

According to the results from Table 5, the LSTM outperforms the ARIMA and Holt-Winters. LSTM and ARIMA has better results than Holt-Winters in terms of RMSE for short sequences.

Performance Matrix Measure for the Sever Log Events

The confusion matrix is used to measure the accuracy and correctness of the model, and it is the matrix of actual and prediction. According to our case study, we have the following categories. **True positive** is the outcome when the model correctly predicts the observed sequence event, such as the warning E9 accurately predict from the observed sequence E8, E10, E9, where E8 and E10 error occurs together, and **True negative** occurs, when the model also confronts the observed sequence. **False positive** is the outcome, when the model incorrectly predicts observed sequence of such as model predict the event even if the sequence does not exist, and **False negative** occurs when model does not confront the observed sequence of events. In this case study, we want to know

Table 5: The RMSE of each predictive model for each time-series, we have for log analysis and we measure the accuracy of predictions for each model.

	LSTM	Holt-Winters	ARIMA
Monitoring Server Logs	1.442	7.830	2.544
Admin server logs	1.523	8.365	2.666
Proxy server logs	1.287	7.256	3.236

Table 6: The outcome of the events in the confusion matrix for LSTM model forecasting results with 82% accuracy.

	Predicted Yes	Predicted No
Actual Yes	27	3
Actual No	6	14

that the predicted sequence confronts observed sequence. We took sample data of 50 observation and forecasting results from LSTM model.

LSTM Model

We have set up the experiments on a local computer and used the Jupyter notebook with Python 3 and Keras libraries for LSTM. We prepare the training and testing data for machine learning models and have used only single LSTM layer. The LSTM is the topology of Neural network algorithm which can hold and learn the observed data and make prediction based on the observed data [21] and does not require stationary data which is the basic requirement of the statistical models. The algorithm takes the dataset needs to be trained, batch size, data fitted size, and neurons. We have fitted the dataset to the model and repeated the learning time five times.

We input the file in one batch because the size of the file is small enough to fit in one batch. The predictive analysis of time series in Figure 3 shows accurate predictive results according to the observed logs for complete data, and in Figure 4 we have shown the predicted events over the 24 hours and forecasted event for next unknown 30 minutes which is marked. LSTM learns from previous data and keeps the record in short-term memory so that it can make the prediction for the future. In this chapter, we consider forecasting time and event separately, however, in future, we will explore to combine the two inputs to make event prediction and forecasting the future events.

ARIMA Model

We have got significant results from ARIMA forecasting model, and we have also calculated. Root mean square error for checking the goodness of the model. We have seasonality in our data, such as more errors and warnings happen in regular business hours rather than weekends and off work hours. In the IT department, some of the batch files run in the night or weekends which often generate the errors. We consider the ARIMA model in order to cross-check the validity of the results. ARIMA (p, d, q) are the critical parameters for predictions and forecasting time series. Figure 6 shows the predictive values of time series and original values from the dataset, and we observe that the prediction of the ARIMA model does not show significant results for long sequences and complete dataset file, however in Figure 7 the forecast is based on short sequence of 30 min window data, which shows promising results.

In ARIMA model, picking up the right parameters for the forecasting is critical for the accurate results, if the parameters are not correct, the results may vary, and the RMSE will be higher for predicted values. PACF is used to determine the value of p and differencing is to determine d and ACF is to determine q.

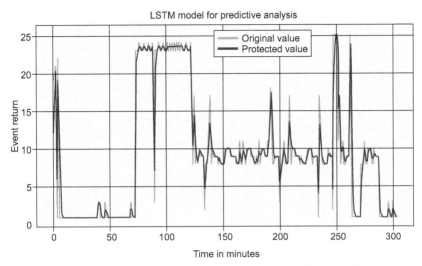

Fig. 3: The predictive results of the LSTM model. The predicted value and original values are approximately close with 1.4 average root mean square error for the complete data.

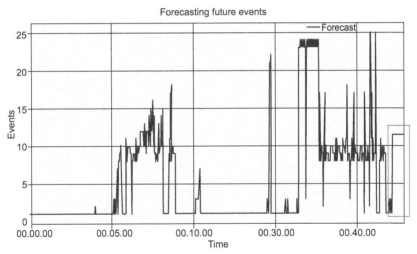

Fig. 4: The forecasted event in the next 30 minutes for future prediction, here we have 24 hours predicted values based on learned data from the history and forecasted the next event.

Holt-Winters Model

Holt-Winter is a triple exponential method, which can predict and forecast the future values based on observed behavior of the time series. This model also accepts three parameters (α,β,γ) such as α determines the weighted average of the points, which decay exponentially; β determines the slope between consecutive points and γ determines the seasonality of the series of the

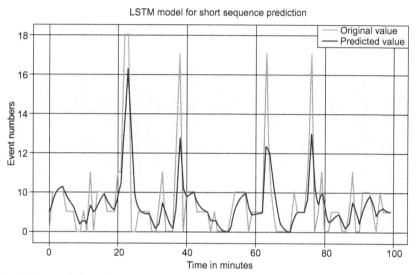

Fig. 5: The prediction of short sequences through the LSTM model. We have also experimented for the short sequence prediction, and LSTM has given promising results in predicting the short sequences. The graph also shows the seasonality in the events every 30 min.

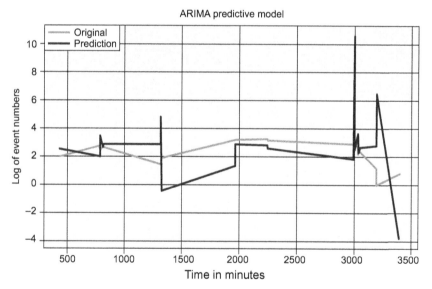

Fig. 6: ARIMA predictive model results for long sequences and data for the complete log file with RMSE of average 2.7.

dataset. We use additive and multiplicative seasonality which in turn reduced the RMSE. Figure 8 shows the results of holt-winter model, by applying triple exponential smoothing.

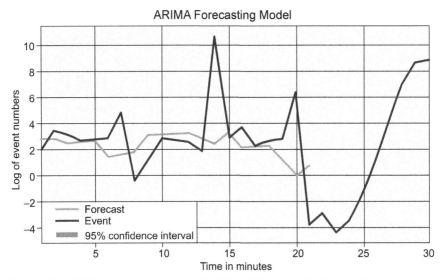

Fig. 7: The ARIMA forecasting results with 95% confidence for the 30 min window and short sequences and shows the original data with forecasted values.

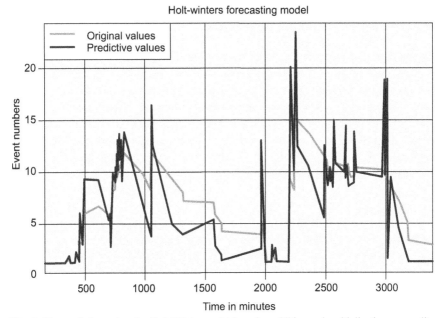

Fig. 8: The predictive values by Holt-Winter model by using additive and multiplicative seasonality with RMSE of 7.8.

6. Discussion and Conclusion

In this chapter, we analyzed log files with the goal of predicting the events and the time of the event. We present the pre-processing technique for feature extraction and also extract the sequence of events by a static window and moving window from the logs. We explore the traditional and modern techniques of the time-series prediction and compare the results, which shows that LSTM outperforms for the prediction of server log data and analyzing the error data. We have proposed the approach to generate log events and the sequence of events from logs and predict the future events based on historical data. To evaluate this approach, we took three different types of log files from various servers, and we have also observed that the ARIMA model works better on a short sequence of events and LSTM model brings better results for shorter and large sequences or large data set. In this chapter, we provide the implementation technique for LSTM, ARIMA, and Holt-Winters and compared the performance of all three models by calculating Root mean square error, which is shown in Table 5. We have visualized the predicted events of the LSTM, ARIMA, and Holt-Winters model, which provides the insight to the data analysts for decision making.

7. Limitations and Future Work

We have compared the log data from three different servers to analyze the results and validate our approach; however, we are also exploring the diverse set of log data from an open source in order to generalize our approach. We use a fixed window of 30 min technique for this approach; if we change the size of the window, it is entirely possible the size of sequences will change which can affect the results. In our future work, we are looking to implement the multi-task learning in deep learning to combine the categorical and numeric data of the lags and make the prediction based on all the selected features.

References

[1] Jiang et al. 2008. An automated approach for abstracting execution logs to execution events. Journal of Software Maintenance and Evolution: Research and Practice 20(4): 19.
[2] Lin et al. 2016. Log clustering based problem identification for online service systems, presented at the Software Engineering Companion (ICSE-C). IEEE/ACM International Conference, Austin, TX, USA.
[3] Hatonen et al. 2008. Local anomaly detection for network system log monitoring. Information Sciences 178(20).
[4] Shang et al. 2013. Assisting developers of big data analytics application when deploying on hadoop clouds, presented at the Software Engineering (ICSE), 2013. 35th International Conference, San Francisco, CA, USA.

[5] Wei et al. 2010. Detecting Large-scale System Problems by Mining Console Logs. Elsevier, 37–44.

[6] He et al. 2016. Experience report: System log analysis for anomaly detection, presented at the Software Reliability Engineering (ISSRE), 2016. IEEE 27th International Symposium, Ottawa, ON, Canada.

[7] Han et al. 2006. Data Mining: Concepts and Techniques. (The Morgan Kaufmann Series in Data Management Systems). Elsevier.

[8] Bovenzi et al. 2014. An os-level framework for anomaly detection in complex software systems. Presented at the IEEE Transactions on Dependable and Secure Computing.

[9] Krishna et al. 2015. Topic modeling of SSH logs using Latent Dirichlet allocation for the application in cyber security. pp. 75–79. *In*: Systems and Information Engineering Design Symposium, IEEE.

[10] Pengtao. 2013. Integrating Document Clustering and Topic Modeling.

[11] Chandola et al. 2009. Anomaly detection: A survey. ACM Computing Surveys (CSUR) 41(3).

[12] Adam and Wei. 2012. Advances and challenges in log analysis. Communications of the ACM, 7.

[13] Python. (March, 2018). https://pypi.python.org/pypi.

[14] Mugume et al. 2016. Comparison of parametric and nonparametric methods for analyzing the bias of a numerical model. April 2016 Modelling and Simulation in Engineering, Volume 2016, Article ID 7530759, 7 pages.

[15] Botchkarev. 2019. Performance metrics (error measures) in machine learning regression, forecasting and prognostics: Properties and typology. Interdisciplinary Journal of Information, Knowledge, and Management 14: 45–79.

[16] Suzgun et al. 2019. On Evaluating the generalization of LSTM Model in formal languages. Proceedings of the Society for Computation in Linguistics (SCiL) 2019, cite as. arXiv:1811.01001v1 [cs.CL].

[17] Suman et al. 2018. Visualization of server log data for detecting abnormal behaviour. IEEE International Conference on Information Reuse and Integration (IRI) pp. 244–247.

[18] Borges et al. 2017. Predicting target events in industrial domain. SpringerLink Books Lecture Notes in Computer Science 10358: 17–31.

[19] Shuyang et al. 2017. Modeling approaches for time series forecasting and anomaly detection. Computer Science Project 2017 Final Report. Available online http://cs229. stanford.edu/proj2017/final-reports/5244275.pdf.

[20] Salfner and Tscirpke. 2008. Error log processing for accurate failure prediction. pp. 4–4. Proceeding WASL'08 Proceedings of the First USENIX Conference on Analysis of System Logs, San Diego, California.

[21] Namin et al. 2018. Forecasting Economins and Financial Time Series: ARIMA vs LSTM. CoRR abs/1803.06386.

[22] Williams and Hoel. 2003. Modeling and forecasting vehicular traffic flow as a seasonal ARIMA process: Theoretical basis and empirical results. Journal of Transportation Engineering 129: 664–672.

[23] Chatfield and Yar. 1988. Holt-Winters forecasting: some practical issues. The Statistician, pp. 129–140.

[24] Brownlee. 2017. How to create an arima model for time series forecasting with python. Available at https://machinelearningmastery.com/arima-for-time-series-forecasting-with-python/.

[25] Brownlee. 2016. Time series prediction with LSTM recurrent neural networks in python with keras. Available at https://machinelearningmastery.com/time-series-prediction-lstm-recurrent-neural-networks-python-keras/.

[26] Kalekar, Rekhi. 2008. Time series forecasting using holt-winters exponential smoothing. School of Information Technology 4329008: 1–13.

[27] Hay et al. 2008. OSSEC host-based intrusion detection guide. Elsevier Science 3 March 2008.

Index